Friction and the Hot Rolling of Steel

Vladimir Panjkovic

CRC Press is an imprint of the
Taylor & Francis Group, an informa business

CRC Press
Taylor & Francis Group
6000 Broken Sound Parkway NW, Suite 300
Boca Raton, FL 33487-2742

First issued in paperback 2017

Version Date: 20140114

ISBN 13: 978-1-138-07717-1 (pbk)
ISBN 13: 978-1-4822-0589-3 (hbk)

Library of Congress Cataloging-in-Publication Data

Panjkovic, Vladimir.
Friction and the hot rolling of steel / Vladimir Panjkovic.
pages cm
Includes bibliographical references and index.
ISBN 978-1-4822-0589-3 (hardback)
1. Rolling (Metal-work) 2. Friction. I. Title.

TS340.P235 2014
671.3'2--dc23 2013048178

Visit the Taylor & Francis Web site at
http://www.taylorandfrancis.com

and the CRC Press Web site at
http://www.crcpress.com

Contents

Section IV Appendices: Technical Details

Preface

One of the greatest obstructions to the mechanical powers of engines proceeds from the friction, or resistance of the parts rubbing on each other; which in general, is greater, or less, as the rubbing parts bear the greater, or less pressure; and yet this obstruction is but little attended to. The theorist makes no allowance on account of friction; and the practical mechanician, who feels the effects, yet, as if unavoidable, seldom takes the trouble of searching for a remedy.

Fitzgerald, 1763

But, however important a part of mechanics this subject may constitute, and however, from its obvious uses, it might have been expected to have claimed a very considerable attention both from the mechanic and philosopher, yet it has, of all the other parts of this branch of natural philosophy, been the most neglected. The law by which the motions of bodies are retarded by friction has never, that I know of, been truly established.

Vince and Shepherd, 1785

From the attention that has hitherto been paid to this important branch of mechanical science, and from the many elaborate dissertations and experiments that have appeared at different periods, it would naturally be concluded, that the subject had been so fully elucidated, as to admit of little if any further investigation: but the diversity of opinions still prevalent among philosophers, and the difficulty of reducing to a satisfactory state the doctrines already advanced, incline me to the opinion that the subject is as yet but imperfectly understood.

Rennie, 1829

... a complete description of its fundamental causes and a single quantitative model—which is generally applicable to any frictional situation—remains elusive.

Blau, 2009

Friction is everywhere, and it affects our lives in both good and bad ways; a car expends 20% of fuel to overcome friction in the engine and drive train, but cannot move on a slippery surface (Burke, 2003). Without friction, we would be slipping and falling instead of walking, and steel strips would never enter the gap between rolls in a steel mill.

The bad aspects of friction seem to be more noticeable. In 1904, Davis described it as a '... highway robber of mechanical energy ... levying tribute on all matter in motion, exerting a retarding influence and requiring power to overcome it.' Friction is associated with wear, and various estimates in the United States, Great Britain and Germany suggest that friction and wear cost many billions of dollars annually (Rabinowicz, 1995; Ludema, 1996; Stachowiak and Batchelor, 2005).

Friction in the hot rolling of steel is particularly important. These days, more than a billion tonnes of steel are produced annually, and most of that steel is hot rolled. Many practical problems of hot rolling are linked to friction: chatter, skidding, excessive rolling force, very high friction of some rolls and the like. These problems prompted the author to study a vast body of literature, only to find out that

1. There are no satisfactory mathematical models of friction in either hot rolling or general engineering.
2. Even the qualitative understanding of it leaves much to be desired.
3. Many laboratory experiments were reported, but the findings and interpretations often contradict each other.

However, the available information can improve the control and understanding of friction in hot strip mills and other industrial plants. First, many problems can be understood if one is aware, at least qualitatively, of the base mechanisms of friction. Second, critical analysis of the literature data, combined with the observations in commercial plants, may explain some contradictions.

The main aim of this book is to present this body of knowledge systematically, and make it available to the wide engineering audience. It is organised in four sections:

Section I (Chapters 1–4), which outlines the history of our understanding of the fundamental causes of friction, from Leonardo da Vinci to the twenty-first century. Understanding of these causes will make the analysis of frictional phenomena in engineering much easier.

Section II (Chapters 5–11) covers the general phenomena relevant to the rolling of metals. These include the impact of roughness and velocity, basics of liquid and solid lubrication, mathematical modelling and the properties of materials that affect friction in steel rolling, such as metals, oxides and carbides.

Section III (Chapters 12–17) connects the theoretical concepts, laboratory-scale observations and phenomena in other areas of science and engineering to the large-scale industrial process of hot rolling. It addresses roll properties, oxidation, wear and chemical composition of rolls and their impact on friction, as well as the evolution of friction over schedules and roll campaigns, and mathematical modelling of friction in hot rolling, with examples from a five-stand, million-tonnes-per-year commercial mill.

Section IV gives technical details, that is, the properties of important species, and interesting diversions, which are presented in appendices.

Finally, some details should be mentioned:

- Wear and lubrication are only considered to the extent relevant to friction in steel rolling.
- The term 'tribology' is often used in this document. Basically, it is the science of friction, wear and lubrication, or as Persson (1999) put it more rigourously, the 'science and technology of interacting surfaces in relative motion'.

References

Blau, P.J. 2009. *Friction Science and Technology,* 2nd ed. Boca Raton, FL: CRC Press.

Burke, S.A. 2003. *Friction Force Microscopy: Seeking New Understanding of Friction from a Nanoscale Perspective*. Unpublished report. McGill University, 28 February.

Davis, W.M. 1904. *Friction and Lubrication. A Hand-Book for Engineers, Mechanics, Superintendents and Managers*. Pittsburgh: Lubrication.

Fitzgerald, K. 1763. A method of lessening the quantity of friction in engines. *Phil. Trans.* 53:139–158.

Ludema, K. 1996. *Friction, Wear, Lubrication*. Boca Raton, FL: CRC Press.

Persson, B.N. 1999. Sliding friction. *Surf. Sci. Rep*. 33:83–119.

Rabinowicz, E. 1995. *Friction and Wear of Materials,* 2nd ed. New York: Wiley-Interscience.

Rennie, G. 1829. Experiments on the friction and abrasion of the surfaces of solids. *Phil. Trans*. 119:143–170.

Stachowiak, G., and A.W. Batchelor. 2005. *Engineering Tribology,* 3rd ed. Boston: Butterworth-Heinemann.

Vince, S., and A. Shepherd. 1785. On the motion of bodies affected by friction. *Phil. Trans*. 75:165–189.

Acknowledgements

This is my first book, and I would like to thank Jonathan Plant and Amber Donley of CRC Press for the assistance given during its preparation, for their courtesy, and for taking stress out of this rather strenuous exercise. The kind permission by the following institutions to reproduce copyrighted images and tables is gratefully acknowledged: The Association for Iron and Steel Technology, EDP Sciences, Elsevier, Fédération Française de l'Acier, Iron and Steel Institute of Japan, Lafayette Photography (Cambridge), The Minerals, Metals and Materials Society, Oxford University Press, Royal Society Publishing, Springer, Taylor & Francis, and Wiley–VCH.

My interest in friction was sparked by enthusiastic discussions with the expert roll technologist, John Steward, followed by involvement in his carefully organised plant trials of novel roll designs and operational conditions. The author was also lucky to have excellent colleagues to exchange ideas with, in particular Gregory Fraser, Ron Gloss, Arthur Liolios, Boris Srkulj, Robert Steward and Peter Wray. I would also like to acknowledge the supply of high-quality photos and very constructive remarks by Dr Mario Boccalini Jr, of the Institute de Pesquisas Tecnologicas, São Paulo, and Prof Amilton Sinatora of the University of São Paulo.

My wife Milica and daughter Tamara showed enormous patience and provided a loving environment during the long hours I spent on my iMac preparing the manuscript, for which I am very grateful. Finally, I would like to thank my parents Olga and Ljubomir for all their support in my quest for knowledge.

About the Author

Vladimir Panjkovic, PhD, grew up in the picturesque historic town Srbobran, in the Serbian province of Vojvodina. He graduated with a degree in electrical engineering from the University of Novi Sad, majoring in control and computer engineering, and earned his PhD in materials science and engineering from the University of New South Wales, for mathematical modelling of novel ironmaking processes. After a brief stint as a research assistant at the University of Novi Sad, he has worked for a quarter century as a research and process engineer in the Australian steel and manufacturing industry. His fields of work include the applications of artificial intelligence to process control, development and deployment of mathematical models of ironmaking processes and steel rolling, analysis of tribological problems in hot strip rolling, and the design and commissioning of thermal equipment. For his work Dr Panjkovic has been awarded the BlueScope Steel Research Excellence Award, the National Project Excellence Award in Automation, Control and Instrumentation, and the John A. Brodie medal in chemical engineering, the latter two by the Institute of Engineers Australia. He has published 7 journal papers, more than 20 conference papers, approximately 40 technical reports, and is commissioned to prepare the chapter on vibrations in steel rolling for the forthcoming *Encyclopaedia of Iron, Steel and Alloys*, to be published by Taylor & Francis in late 2015.

Section I

History of Friction: From da Vinci to Now

1

Early Studies of Friction

That there is a Loss of Force in the working of Engines on account of the Rubbing or Friction of their Parts, has been observ'd by most Writers of Mechanics; but that Friction has not been enough consider'd by them ... Projectors contrive new Machines (new to them, tho' perhaps describ'd in old Books, formerly practiced and then disus'd and forgot) which they suppose will perform much more than they have seen done with the same Power; because they allow too little for Friction. Full of this they go to the Charge of 70 or 80 l. for a Patent for their new Invention; then divide it into Shares, and draw in Persons more ignorant than themselves to contribute towards this (suppos'd advantageous) Undertaking; till after a great deal of Time and Money wasted, they find their own Engine worse than others which they hoped by many degrees to excel. This has been very much the Practice for these last twenty Years : For tho some Projectors have been altogether Knaves, yet the greatest part have first deceiv'd themselves; and those who are really deceiv'd, by their eagerness and earnestness most easily deceive and draw in others. For this reason, I thought it would be of Use to the Publick, to give as full an Account of friction, as I possibly could gather from the Experiments made by others (especially the Members of the Royal Academy at Paris) and my own Experiments and Observations.

Desaguliers, 1745

Those who cannot remember the past are condemned to repeat it.

Santayana, 1905

Solomon saith, There is no new thing upon the earth. So that as Plato had an imagination, That all knowledge was but remembrance; so Solomon giveth his sentence, That all novelty is but oblivion.

Bacon, 1787

When studying a subject, it is prudent to study its history first, even briefly, to avoid the reinvention of wheel. The history of tribology contains many examples of sound ideas that were forgotten or ignored, and rediscovered much later:

1. Leonardo da Vinci proposed the two basic laws of friction in the beginning of the sixteenth century, but they were rediscovered by Amontons almost two centuries later.
2. Robert Hooke proposed in 1685 that deformation and adhesion are the primary causes of friction, which is consistent with modern views. The adhesion was refuted by Leslie in 1804, and revived more than a century later.
3. Guillaume Amontons in 1699 represented the elastic contacts between surfaces with springs or bristles. This concept is now widely used in the friction models devised by control engineers.

This history starts with addressing da Vinci, the first known person to study friction scientifically, and the works of Hooke and Amontons in the seventeenth century. The next section is dedicated to developments in the eighteenth century, followed by the section on the revolution in liquid lubrication and the research of dry friction in the nineteenth century. The fourth section is about the progress of tribology in the twentieth century and beyond. It includes the Stribeck curve, the assertion of Bowden and Tabor that adhesion and ploughing are the key causes of friction, studies at the atomic scale, and the application of thermodynamics to the calculation of the coefficient of friction (COF, μ) in hot rolling.

1.1 Leonardo da Vinci

Leonardo da Vinci (1452–1519, Figure 1.1) was the first known person to conduct systematic experiments with friction and summarise observations as laws. These laws were quoted by Dowson (1998) as follows (the literal translation from Italian is in italics):

1. The force of friction is directly proportional to load (*friction produces double the amount of effort if the weight be doubled*).
2. The friction is independent of the apparent contact area (*friction made by the same weight will be of equal resistance at the beginning of its movement although the contact may be of different breadths and lengths*).

He noted that for smooth surfaces 'every frictional body has a resistance of friction equal to one-quarter of its weight'; that is, μ is 0.25, which is well within the range encountered in practice.

Amontons conducted the next systematic studies almost two centuries later, and literally rediscovered his laws. Although the contributions of da Vinci and Amontons to tribology are well known, they are rarely

FIGURE 1.1
Leonardo's self-portrait. (Reprinted from da Vinci, http://upload.wikimedia.org/wikipedia/commons/b/ba/Leonardo_self.jpg. Accessed May 10, 2013. With permission.)

appreciated, with the notable exception of Dowson, that, before Amontons, Hooke reported profound ideas about the nature and control of friction. The works of these two pioneers are described in more detail below.

1.2 Robert Hooke

For reasons widely debated, Hooke (1635–1703) was much maligned by his contemporaries and posterity, although the greatest gossiper of that day, Aubrey, recorded that '... he is of prodigious inventive head; so he is a person of great vertue and goodness ... certainly the greatest mechanick this day in the world' (Anon., 1813). This polymath contributed to many areas of science, and at the age of 30 published the first scientific bestseller, *Micrographia*. In 1685, deeply impressed by Stevin's sailing chariot, a sailing boat on wheels devised for Prince Maurice of Orange, he proposed the following notions (Hooke, 1685):

1. The rolling friction is influenced by deformation and adhesion, which is a modern view (see Section 4.4): 'Next, we are to consider, what Impediment to its Motion, a Wheel, thus roll'd upon a Floor,

receives from that Floor. ... The first and chiefest, is the yielding, or opening of that Floor, by the Weight of the Wheel ...; and the second, is the sticking and adhering of the Parts of it to the Wheel'.

2. Adhesion is analysed in detail: 'The Second Impediment it receives from a Floor, or Way, is the sticking and adhering of the Parts of the Way to it ... there is a new Force requisite to pull it off, or raise the hinder Part of the Wheel from the Floor, or Way, to which it sticks. ... The force of adhesion depends on surface properties: ... the harder the Ways are, the less Impediment they give to the Motion of Carriages. ...'
3. The term 'friction' was used for the first time in its modern mechanical meaning, as a phenomenon hindering motion (see Appendix A): '... because the gudgeons, halving the weight, may be made very much smaller, and so will not cause a tenth part of the *friction* which is necessary in the other way'.
4. Amontons proposed 14 years later that the friction is caused by the asperities of one surface climbing up over those on the opposing surface (Amontons, 1699). Hooke had discounted this well before the better known refutation by Leslie (1804; see also Section 3.1):

   > ... for, if the Floor be perfectly hard (as also the Parts of the Wheel) tho' it be very unequal, yet is there little or no Loss, or considerable Impediment to be accounted for; for whatever Force is lost, in raising or making a Wheel pass over a Rub, is gain'd again by the Wheel's descending from that Rub, in the same Nature as a Ship on the Sea is promoted by the descending down of a Wave, as much as impeded by its ascending, or a Pendulum is promoted by its Descent, as much as impeded by its Ascent.

5. Practical advice is given to reduce friction:

   > The less rubbing there be of the axle, the better it is for this effect; upon which account, steel axes, and bell-metal sockets, are much better than wood, clamped, or shod with iron; and gudgeons of hardened steel, running in bell-metal sockets, yet much better, if there be provision made to keep out dust and dirt, and consistently to supply and feed them with oil, to keep them from eating one another. ...

As Dowson observed, Hooke recommended the use of soft metal bearings to reduce friction. He anticipated the ideas proposed by Bowden and Tabor (Section 4.4) in the mid-twentieth century, without knowing the concept of shear strength.

Larsen-Basse (1992) proposed that regarding the causes of friction, there were two early schools. The French school was promoted by Amontons, and later Coulomb, and emphasised the mechanical (elastic) interaction of surface roughness and asperities. On the other hand, the English school, represented primarily by French-born Desaguliers, advocated 'cohesion', or adhesion between the materials. Even before the French school was established, Hooke refuted it, and preceded the English school by about a half century.

1.3 Guillaume Amontons

Da Vinci's laws were unnoticed or forgotten, to be rediscovered via the systematic experiments by Amontons (1663–1705), who summarised the findings as follows (1699):

1. The resistance caused by friction increases or diminishes in proportion to pressure, the magnitude of which is larger or smaller depending on whether the area of rubbing surfaces is bigger or smaller.
2. The resistance caused by rubbing is similar for iron, copper, lead and wood combined in any manner, if the surfaces are coated with old pork fat.
3. The resistance is about one-third of the load, suggesting $\mu \sim 0.33$.
4. The resistance between the rubbing bodies depends in a complex way on normal pressure, time and sliding speed.

Amontons' laws are derived from the first statement, and it can be seen that they are practically identical to da Vinci's rules:

1. The force of friction is directly proportional to the applied load.
2. The force of friction is independent of the apparent area of contact.

Amontons contended that friction on hard surfaces is associated with the force required to lift asperities of one surface over those of another, in a movement along an inclined plane. On softer surfaces, there could be an elastic component of friction, and it was represented with elastic springs as shown in Figure 1.2. This representation has been widely used in the friction models developed by control engineers (Canudas-de-Wit et al., 1995).

According to Kragelsky and Shchedrov (1956), these findings were received with some scepticism by the French Academy. Philippe de la Hire (1640–1718) was commissioned to verify them, and they were confirmed by his experiments. De la Hire developed his theory:

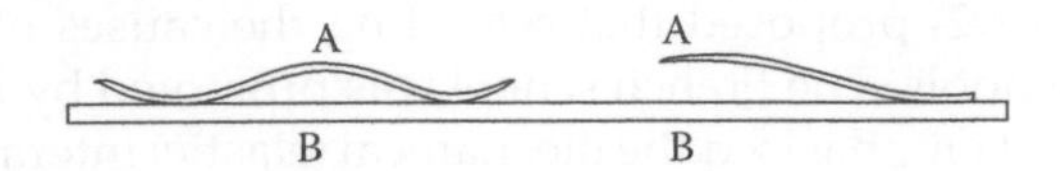

FIGURE 1.2
Representation of softer surfaces with elastic springs. (Reprinted from G. Amontons, *Histoire de L'Académie Royale des Sciences*: 206–227, 1699. With permission.)

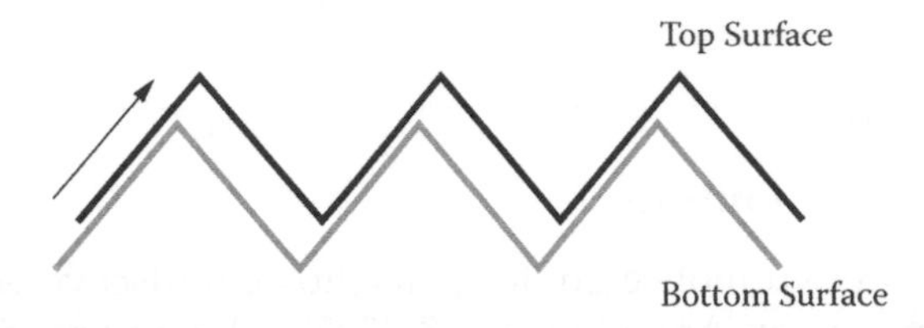

FIGURE 1.3
Climbing of asperities at the top surface over the asperities at the bottom surface.

1. Friction is caused by the interlocking of asperities, which are either elastic or hard.
2. The elastic ones are bent like springs, and the more bent they are, the larger the friction. At given pressure, the bending is inversely proportional to the number of springs, and that's why friction does not depend on surface area.
3. If asperities are hard, force is required to lift the asperities of one surface over those on the opposite surface (Figure 1.3). The friction is then proportional to pressure.

De la Hire also envisaged a case when the friction depends on the contact surface area. That occurs when the asperities are broken, that is, snipped during motion. The resistance to motion is then proportional to the number of broken asperities, that is, to the surface area.

Generally, Amontons' laws hold in many practical cases. However, Ringlein and Robbins (2004) quoted the example of sticky tape, which exhibits friction without load. Also, with sticky and compliant objects, friction increases with the contact area.

References

Amontons, G. 1699. De la resistance cause'e dans les machines. *Histoire de L'Académie Royale des Sciences*: 206–227.

Anon. 1813. *Letters written by eminent persons in the seventeenth and eighteenth centuries: to which are added, Hearne's journeys to Reading, and to Whaddon Hall, the seat of Browne Willis, Esq. and Lives of eminent men, by John Aubrey, Esq.* London: Longman, Hurst, Rees, Orme, and Brown, and Oxford: Munday and Slatter.

Bacon, F. 1787. *The essays of Francis Bacon, Baron of Verulam, Viscount St. Alban, and Lord High Chancellor of England, on Civil, Moral, Literary and Political Subjects, Together with the Life of That Celebrated Writer, Vol. 1.* London: Logographic Press. Originally published in Latin in 1625.

Canudas-de-Wit, C., H. Olsson, K.J. Åström et al. 1995. A new model for control of systems with friction. *IEEE Trans. AC* 40:419–425.

Desaguliers, J.T. 1745. *A Course of Experimental Philosophy, Vol. I,* 2nd ed. London: Innys, Longman, Shewell, Hitch and Senex.

Dowson, D. 1998. *History of Tribology,* 2nd ed. London and Bury St Edmunds: Professional Engineering Publishing Ltd.

Hooke, R. 1685. Discourse of carriages before the Royal Society, on Feb. 25. 1685. With a description of Stevin's Sailing Chariot, made for the Prince of Orange. In *Philosophical Experiments and Observations of the Late Eminent Dr. Robert Hooke,* 150–167, 1726. London: W. Derham and J. Innys.

Kragelsky, I.V. and V.S. Shchedrov. 1956. *Development of the Science of Friction: Dry Friction*. Moscow: Academy of Sciences of USSR.

Larsen-Basse, J. 1992. Basic theory of solid friction. In *Friction, Lubrication, and Wear Technology, ASM Handbook 18*, 27–38. Ohio: American Society for Metals.

Leslie, J. 1804. *An Experimental Inquiry into the Nature and Propagation of Heat.* Edinburgh: Mawman.

Ringlein J., and M.O. Robbins. 2004. Understanding and illustrating the atomic origins of friction. *Am. J. Physics* 72:884–891.

Santayana, G. 1905. *The Life of Reason, or, the Phases of Human Progress.* New York: Scribner's, and London: Constable.

da Vinci, L. http://upload.wikimedia.org/wikipedia/commons/b/ba/Leonardo_self.jpg (accessed May 10, 2013. With permission.)

2

Eighteenth Century

This century saw a surge in the studies of friction. Many notable scientists in Western Europe (France, Great Britain, Holland and German-speaking lands) conducted experiments and/or proposed the causes of friction. The development of science started in earnest in Russia, and friction attracted significant attention.

2.1 France: Parent, Camus, Bélidor

Between Amontons and Coulomb, there were three French tribologists of some note:

1. In 1700, Antoine Parent (1666–1716) reported observations that, with light lubrication, the COF is ~0.33 for iron, lead, copper and wood. He confirmed Amontons' results, but considered that speed does not affect friction. However, he proposed that the COF may differ slightly between various materials (Kragelsky and Shchedrov, 1956).
2. François Joseph des Camus (1672–1732) published in 1724 the results of his extensive experiments, and pointed out that the COF depends on the physical properties of surfaces, that is, whether they are dry, wet, or lubricated. The observed range (0.15–0.45) was much wider than reported by anyone earlier. He also argued that the COF decreases with increasing normal load, and, surprisingly, that lubrication increases friction. Nevertheless, he recommended lubrication, to reduce wear and make sliding smoother (Kragelsky and Shchedrov, 1956).
3. Bernard Forest de Bélidor (1698–1761) is credited to be the first to apply calculus to engineering problems (Day and McNeil, 1996). He published a seminal book on hydraulics, noted both for the depth of knowledge, and for the beautiful illustrations (Figure 2.1). In the book, he represented the surface roughness by the arrays of spheres, and calculated the force required to pull one layer of spheres over another (Bélidor, 1737). The estimated COF was ~0.35.

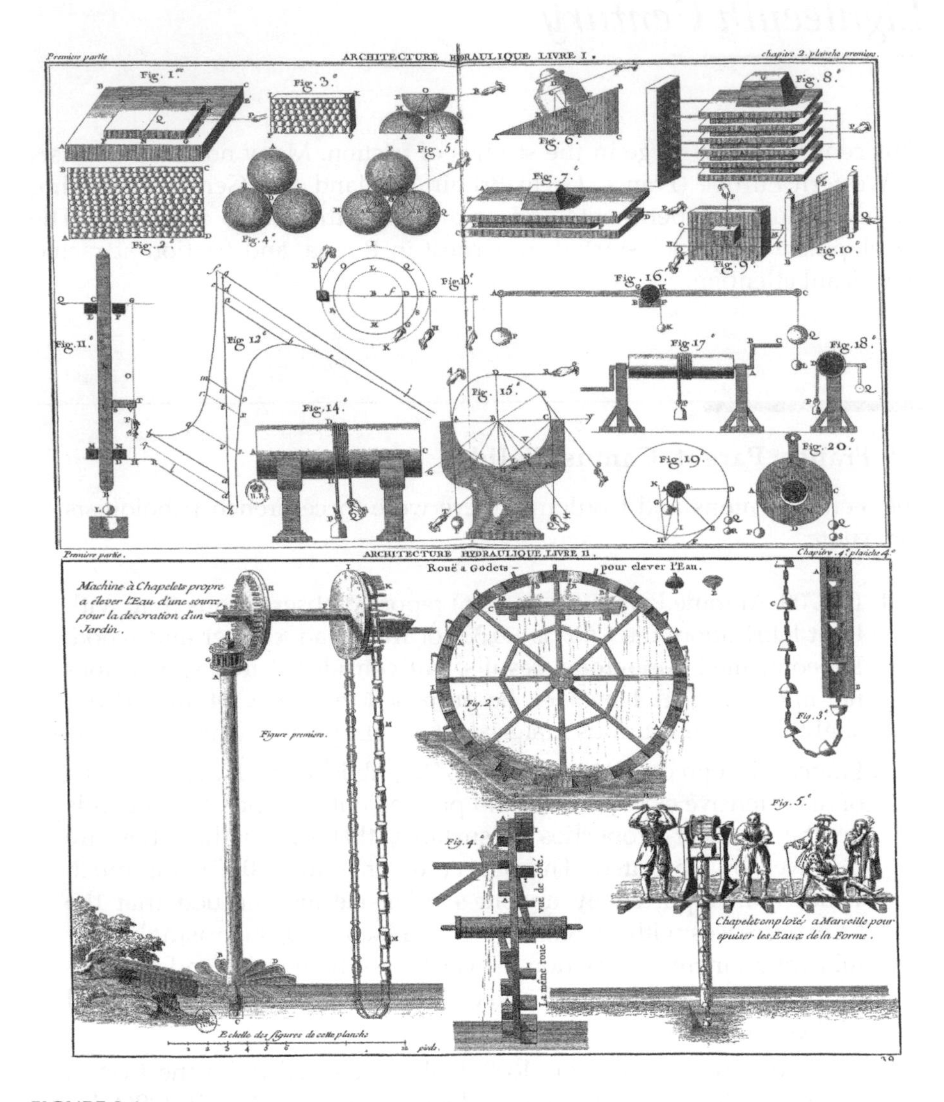

FIGURE 2.1
The top plate shows the COF derivation, and the bottom plate the admirable aesthetics of Bélidor's illustrations. (Reprinted from B.F. de. Bélidor, *Architecture hydraulique, premiere partie*, Paris: C.-A. Jombert, 1737. With permission.)

2.2 German-Speaking Lands: Leibniz, Leupold, Euler

Given their penchant for smart machinery, it is not surprising that Germans became involved in the early studies of friction, as shown by Kragelsky and Shchedrov (1956):

1. Gottfried Wilhelm von Leibniz (1646–1716) briefly delved into friction. He argued that the COF is not constant, but depends on the physical properties of surfaces in contact, and pointed out that the friction of rolling is smaller than the friction of sliding*.
2. Jacob Leupold (1674–1727) was involved in the design of machines. He confirmed the findings of Amontons and obtained a COF of 1/3 for dry wood. Compared to the dry case, the COF roughly doubled with kerosene as a lubricant, and was somewhat smaller for a soaped surface. He also questioned the constancy of $\mu = 1/3$, arguing that it depends on roughness, the properties of rubbing surfaces, and the shape of asperities.
3. Leonhard Euler (1707–1783) introduced the symbol μ for COF in 1748 (Euler, 1750a,b). Also, he concluded by theoretical reasoning that kinetic friction is smaller than static friction†. As did French scientists, Euler explained friction via the climbing of asperities of one body over the asperities on the opposing surface.

2.3 Russia

The treatment of friction in Russia illustrates the importance of its studies in the eighteenth century. A few scientists mentioned here were involved, directly or indirectly.

1. Peter the Great‡ established the Russian Academy of Sciences in 1724 as the Saint Petersburg Academy of Sciences. Leibniz was the key advisor in this undertaking.

* Seireg (1998) claims that Themistius in the fourth century BC had observed that friction is much smaller in rolling than in sliding. However, Themistius lived in the fourth century AD. See also the next footnote.

† Cotterell and Kamminga (1992) argued that Themistius had said that *kinetic* friction is smaller than *static* friction. This is supported by Hecht (2003), who attributed to Themistius the saying: 'Generally, it is easier to further the motion of a moving body than to move a body at rest'.

‡ Peter the Great was technically minded and a science enthusiast. As a young potentate he visited Greenwich and Oxford, studied the city building in Manchester, and inspected shipbuilding in England and the Netherlands. Allegedly, he worked as a carpenter in the largest shipyard of the day, that of the Dutch East India Company, to gain hands-on experience.

2. The czar himself noted the experiments conducted by Leupold, appreciated his book *Theatrum machinarum generale*, and funded some of his research.
3. The czar hired Bélidor to tutor his protégé, Abram Petrovich Hannibal.
4. Pieter van Musschenbroek (Section 2.5) was a member of the academy.
5. The first study of friction in the academy was conducted by German scientist Georg Bernhard Bilfinger (1693–1750), who established that the COF is 0.25.
6. Peter the Great invited distinguished foreigners into the academy, and requested that each of them take two Russian apprentices. The first Russian tribologist of some note, Semyon Kirilovich Kotel'nikov (1723–1806)*, was Euler's apprentice.

2.4 Desaguliers and the Concept of Adhesion

John Theophilus Desaguliers (1683–1744) was an enthusiastic experimenter and promoter of science. He came from France as a Huguenot refugee at the age of 11. He graduated from Oxford, was ordained as a priest of the Church of England, and was such a supporter of Newton that he was called 'more Newtonian than Newton' (Albree and Brown, 2009). He even published a poem: 'The Newtonian System of the World, the Best Model of Government' (Baillon, 2004). Unsurprisingly, Newton was the godfather of his third child.

Desaguliers conducted many experiments and concluded that COF is similar for combinations of lubricated wood, iron, lead and brass (Kragelsky and Shchedrov, 1956). It is likely that the lubricant was thick enough to separate the surfaces and make their properties unimportant for COF. He proposed the method of the calculation of friction force for a system of three pulleys, although it was far from practical (Desaguliers, 1731). More important, he believed that adhesion is the prime cause of friction (1745): '… yet it is found by experience that the flat Surfaces of Metals or other Bodies may be so far polish'd as to increase Friction; and this is a mechanical Paradox; but the Reason will appear when we consider that the Attraction of Cohesion

* Borodich and Keer (2005) argued: 'Using Kotelnikov notation … the coefficient of friction is denoted by μ". Dowson (1998) believed that '… he clearly introduced the concept of a "coefficient of friction", without apparently using this terminology.' He wrote: '… If we denote the friction content F and the applied force P as unknowns, in the ratio μ: 1, then friction $F = \mu P$".' Blau (2001) cited Dowson on this, but (a) Euler introduced both the symbol and the relationship $F = \mu P$ in 1748 and 1749 (Euler, 1750a,b), whereas Borodich and Keer quoted Kotel'nikov's book from 1774; (b) Kragelsky and Shchedrov (1956) quoted the same paragraph as Dowson, but merely commented that Kotel'nikov suggested that the COF may depend on the smoothness of surfaces.

becomes sensible as we bring the Surfaces of Bodies nearer and nearer to Contact.'

Then he applied his earlier observations of the cohesion of lead balls (1724), and proposed that adhesion may be an important cause of friction, in addition to roughness. In those experiments, tops were cut off two lead balls, weighing one and two pounds, and the balls were pressed together at the flat surfaces. The adhesion was so strong, that they would not separate when the smaller one was lifted by hand; the weight exceeding 16 pounds had to be fastened to the larger one to make it fall off. He disagreed with Euler, contending that friction decreases with increasing roughness; smaller roughness ensures better contact between surfaces, which increases adhesion.

Kragelsky and Shchedrov credited Desaguliers as the progenitor of the molecular theory of friction (Dowson was less enthusiastic), but sadly commented that his theory had not been taken seriously then. They also praised his book for its lively style, readable presentation, use of illustrations and experiments, and so on. The presentation was said to have a somewhat 'naïve manner' (Figure 2.2), but it was pointed out that, from both theoretical and practical views, the book was well ahead of contemporaries. Theoretically, it considered molecular mechanisms; practically, experiments were made to quantify the force of friction.

2.5 Pieter van Musschenbroek

As a tribologist, Dutch van Musschenbroek (1692–1761) is best remembered for friction machines that generate electricity (another beautiful drawing, Figure 2.3). He is also credited by some with the first capacitor, the Leyden jar. Dowson credited him for using the concept of bristles to describe the elastic contact points, and Kragelsky and Shchedrov claim that he adhered to the theory of interlocking asperities.

Van Musschenbroek summarized his studies in friction in a posthumous book (1762). Dowson ignored this work, and Kragelsky and Shchedrov claimed that '... almost all his findings were found wrong.' This judgment seems harsh, inasmuch as some statements, quoted by Kragelsky and Shchedrov themselves, make sense, such as that COF differs between various materials, and that identical surfaces have higher friction than different ones, because asperities coincide by shape and dimension, enabling a tight contact. In the extreme case of identically spaced asperities (Figure 2.4a), there is contact over the whole apparent area. Incommensurable surfaces touch only at the peaks of asperities (Figure 2.4b), which significantly reduces the contact area.

Other statements are valid under specific conditions, for example, that friction grows with speed. In liquid lubrication, this is valid in a certain range of the Stribeck curve (Chapter 6), and holds in some cases with dry friction

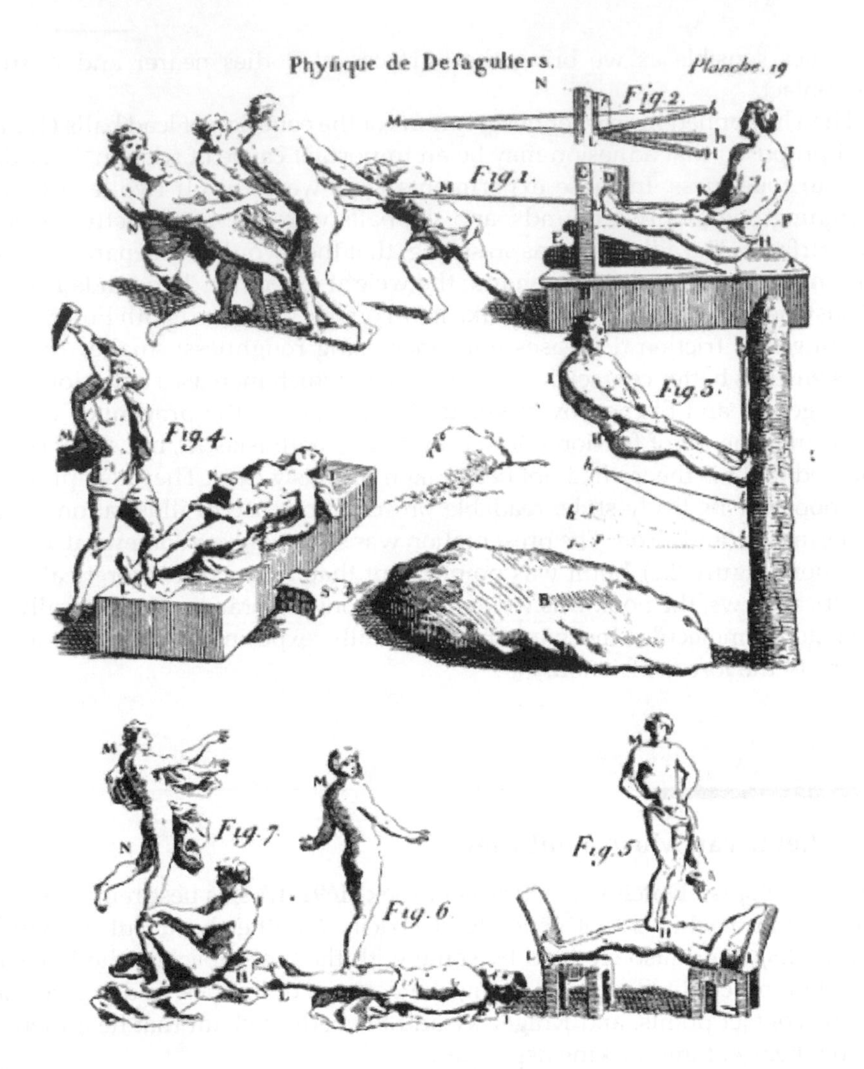

FIGURE 2.2
Drawings from a book on experimental philosophy. Kragelsky and Shchedrov found some of them hilarious, such as the person climbing a pole (marked Fig. 3) who used a quilt with feathers (the pile marked 'B') as a precaution in case of a fall. (Reprinted from J.T. Desaguliers, *Cours de physique expérimentale*, Paris: Rollin & Jombert, 1751. With permission.)

(Section 7.1). However, some statements are questionable indeed. He asserted that friction force depends on surface area. Convolutedly, he claimed that each contact between two bodies is characterised by the surface area corresponding to the 'minimum friction': changing that surface area always leads to increasing friction force. Also, his assertion that lubrication of metallic surfaces is particularly effective at high velocities does not hold for the hydrodynamic lubrication regime (see Chapter 6).

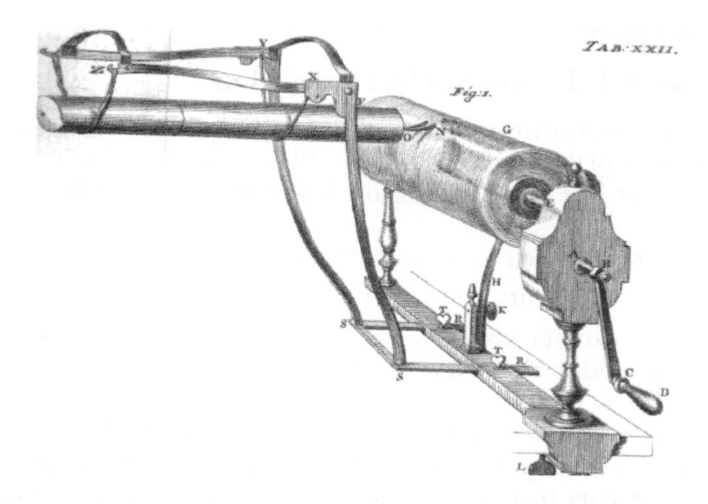

FIGURE 2.3
Generation of electricity by friction. (Reprinted from P. van Musschenbroek, *Introduction ad philosophiam naturalem, Vol. 1,* Leiden: S. et J. Luchtmans, 1762. With permission.)

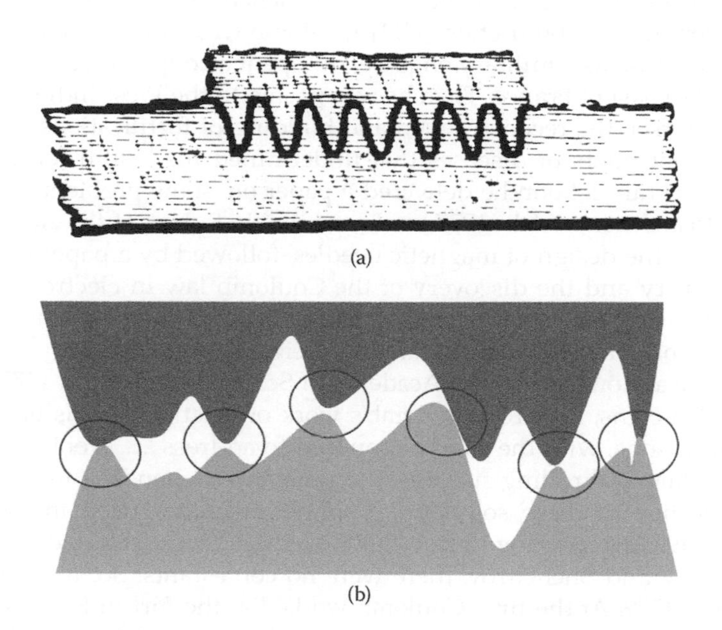

FIGURE 2.4
(a) Commensurable surfaces where asperities touch along their full surface (Reprinted from C.A. Coulomb, *Théorie des machines simples, en ayant égard au frottement de leurs parties et a la roideur des cordages*, Paris: Bachelier, 1821. With permission.) (b) Incommensurable surfaces, with the contacts only at the peaks of asperities, marked by circles.

2.6 Coulomb: Life, and Studies of Friction

Kragelsky and Shchedrov considered Charles Augustin de Coulomb (1736–1806) the creator of the science of friction. They summarised it by saying that before him there had been three major steps in the science of friction:

1. Introduction of the coefficient of friction
2. Discovery of the difference in COF between various materials
3. Observation of the impact of various 'constructive' parameters (presumably modifiable surface conditions) on COF

Before him, researchers simply had noted the impact of various parameters on friction. Coulomb was the first to understand that impact and apply it to his experiments, thereby creating the experimental conditions similar to those in real life, obtaining results applicable in practice.

2.6.1 Life and Motivation

The story below is based on Kragelsky and Shchedrov (1956), Dowson (1998), and Coulomb's book on friction (1821). Coulomb was a man of many talents. After graduating as a military engineer, he spent a couple of years mapping the Atlantic coast of France. Then he was posted to the West Indies for eight years to oversee the reconstruction and building of fortifications. Troubled with frequent bouts of diseases that undermined his health for good, he returned to France. Soon he delivered a paper on 'some problems in statics relating to architecture'. In 1777 he shared the first prize of the Academy of Sciences, for the design of magnetic needles, followed by a paper on dredging machinery and the discovery of the Coulomb law in electrostatics. He made significant contributions to the design of retaining walls, and played important roles in the standardization of weights and measures.

The prize announced by the Academy of Sciences in Paris in 1777, worth 1000 golden coins, inspired Coulomb's work on friction. It was offered for practical reasons, with the essays supposed to address '... problems of friction of sliding and rolling surfaces, the resistance to bending in cords, and the application of these solutions to simple machines used in the navy' (Dowson, 1998). According to Dowson, no winners were selected; according to Kragelsky and Shchedrov, there were no contestants. So, the prize was doubled in 1779. At the time Coulomb worked at the fort in Rochefort, and was well aware of these problems. The local commander allowed him to conduct his friction research, even assigning him two assistants. Coulomb's winning essay consisted of two parts, the first on the sliding friction on planes, and the second on the stiffness and friction of ropes.

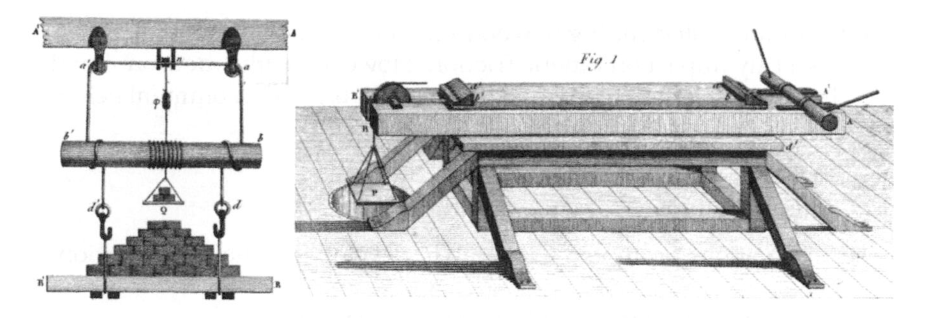

FIGURE 2.5
Some of the equipment used by Coulomb. (Reprinted from C.A. Coulomb, *Théorie des machines simples, en ayant égard au frottement de leurs parties et a la roideur des cordages*, Paris: Bachelier, 1821. With permission.)

2.6.2 Experiments and Observations

Experiments involved several types of timber, and iron and brass; both dry friction and lubrication (with water, olive oil, tallow, axle grease and soot) were studied. Some of his equipment is shown in Figure 2.5. Coulomb concluded that four factors determine friction at the onset of sliding: (a) the nature of surfaces and lubrication, (b) the length of surfaces, (c) pressure, and (d) time passed since the surfaces were joined. He also suggested the fifth cause, humidity of surrounding air, inasmuch as molecules of water could form a thin lubrication layer*. Coulomb, however, restricted this to a conjecture, because he did not notice the impact of humidity on the results of his experiments.

There were other interesting observations:

1. For wood sliding on wood, the friction force is proportional to load at any speed, but kinetic friction is much lower than static friction.
2. For unlubricated sliding of metal on metal, the friction force is proportional to load, and there is no difference between static and kinetic friction.
3. During dry sliding of metal on wood, the static friction increases very slowly with the time of repose and it may take days to reach steady-state.
4. For the metal-on-metal sliding, the steady-state is attained almost instantaneously, whereas for the wood-on-wood it takes a minute or two.

---

* It was noticed during WWII that the flying of aircraft at high altitudes created excessive wear in the carbon brushes of electrical generators. The main cause was the reduced air humidity. Experiments subsequently demonstrated that carbon lubricates well in the presence of moisture, lowering friction and wear (Buckley, 1985).

5. For unlubricated sliding of wood-on-wood or metal-on-metal, speed has a tiny impact on kinetic friction. However, with wood-on-metal the kinetic friction increases with speed (NB: see the comment below on Coulomb's law of friction).

2.6.3 Interpretations

These observations prompted Coulomb to develop this theory of friction.

1. Friction is caused by the interlocking of asperities.
2. The impact of adhesion must be minor, otherwise friction would be proportional to the surface area of contact, because the increased number of contact points would increase adhesion. This view was proven wrong only after the measurements of the real area of contact (see Section 4.4).
3. The wood surface was covered with elastic bristles. The bristles would penetrate each other, and it would take some time for the penetration to settle, hence the impact of the time of repose. With wood-on-metal sliding, bristles of wood tend gradually to fill the cavities between the hard globular asperities on the metal surface. Longer repose time means more thorough filling of cavities with bristles.
4. Once the tangential force is applied, the bristles would start slipping out of the mesh. The fibres are bent under a certain angle, determined by the bristle size, and they form an inclined plane. The angle determines the size of frictional resistance.
5. Once the sliding starts, bristles fold, and the slope of inclined planes decreases, hence the kinetic friction of fibrous substances is smaller than their static friction.
6. The surface of metals is covered with small rigid globules, and there is no significant difference between the static and kinetic friction, inasmuch as those asperities do not fold.

Some authors attribute to Coulomb the third law of friction (Canudas-de-Wit et al., 1995, e.g.), that is, the independence of friction on speed. Coulomb did not propose this law, and actually observed the opposite, which he interpreted as follows:

1. At smaller load, the bending of bristles dominates friction. Higher speed means that more bristles have to be bent in unit time, hence friction increases with speed.
2. At high pressures, bristles do not penetrate each other much, because they get crumpled and crushed, and the higher the velocity, less time they have for penetration, hence friction decreases with speed.

Coulomb also considered what is now known as boundary lubrication. He observed that with a thin coat of lard, the dependence of friction on speed disappeared for the metal–wood contact. He believed that lubricant fills cavities on the metal surface, and smooths the surface. Regarding wood, the unguent glues the bristles together, presumably reducing their capability for interpenetration, and would also reduce their loss of elasticity.

References

Albree, J., and S.H. Brown. 2009. A valuable monument of mathematical genius: *The Ladies' Diary* (1704–1840). *Historia Mathematica* 36:10–47.

Baillon, J.-F. 2004. Early eighteenth-century Newtonians: The Huguenot contribution. *Stud. Hist. Phil. Sci.* 35:533–548.

Bélidor, B.F. de. 1737. *Architecture hydraulique, premiere partie*. Paris: C.-A. Jombert.

Blau, P.J. 2001. The significance and use of the friction coefficient. *Tribol. Int.* 34: 585–591.

Borodich, F.M., and L.M. Keer. 2005. Modeling effects of gas adsorption and removal on friction during sliding along diamond-like carbon films. *Thin Solid Films* 476:108–117.

Buckley, D.H. 1985. Tribology. In *Tribology: The Story of Lubrication and Wear*, 3–20. NASA TM/101430.

Canudas-de-Wit, C., H. Olsson, K.J. Åström et al. 1995. A new model for control of systems with friction. *IEEE Trans. AC* 40:419–425.

Cotterell, B., and J. Kamminga. 1992. *Mechanics of Pre-Industrial Technology*. Cambridge: Cambridge University Press.

Coulomb, C.A. 1821. *Théorie des machines simples, en ayant égard au frottement de leurs parties et a la roideur des cordages*. Paris: Bachelier.

Day, L., and I. McNeil (Eds.). 1996. *Biographical Dictionary of the History of Technology*. London: Routledge.

Desaguliers, J.T. 1724. Some experiments concerning the cohesion of lead. *Phil. Trans.* 33:345–347.

Desaguliers, J.T. 1731. An experiment to show that the friction of the several parts in a compound engine, may be reduced to calculation. *Phil. Trans.* 37:292–293.

Desaguliers, J.T. 1745. *A Course of Experimental Philosophy, Vol. I,* 2nd ed. London: Innys, Longman, Shewell, Hitch and Senex.

Desaguliers, J.T. 1751. *Cours de physique expérimentale*. Paris: Rollin & Jombert.

Dowson, D. 1998. *History of Tribology,* 2nd ed. London and Bury St Edmunds: Professional Engineering.

Euler, L. 1750a. Sur la frottement des corps solides. *Memoires de l'academie des sciences de Berlin* 4:122–132.

Euler, L. 1750b. Sur la diminution de la resitance du frottement. *Memoires de l'academie des sciences de Berlin* 4:133–148.

Hecht, E. 2003. *Physics: Algebra/Trig*. Stamford: Cengage Learning.

Kragelsky, I.V. and V.S. Shchedrov. 1956. *Development of the Science of Friction: Dry Friction*. Moscow: Academy of Sciences of USSR.
Musschenbroek, P., van. 1762. *Introduction ad philosophiam naturalem, Vol. 1*. Leiden: S. et J. Luchtmans.
Seireg, A.A. 1998. *Friction and Lubrication in Mechanical Design*. New York: Marcel Dekker.

3

Nineteenth Century

There was a calm in the studies of dry friction in the nineteenth century. Apart from Leslie's thoughts on the causes of friction, Rennie's thorough experiments, and the initial work on the friction–velocity dependence, the rest was mainly the systematisation of what had been known already. On the other hand, there was a rapid expansion of liquid lubrication, with significant research conducted by Hirn, Petrov and Reynolds.

3.1 Dry Friction

3.1.1 John Leslie

Leslie (1766–1832) was a professor of mathematics and physics at the University of Edinburgh, and made significant contributions to the studies of heat*. In 1804, he refuted the contemporary theories of friction (Leslie, 1804):

1. *Interlocking of asperities.* Leslie observed that friction does not decrease when the surfaces are highly polished; that is, their asperities were diminished: 'By removing the visible asperities from the surfaces of bodies, their mutual attrition is diminished. But any higher polish than what merely prevents the grinding and abrasion of the protuberant particles, has no material effect in reducing the measure of Friction'†.
2. *Climbing of asperities over each other.* Amontons, Euler and Coulomb proposed that friction is due to the energy required for the asperities of one body to climb over those on the opposite surface. However, Leslie noted that the energy expended on the climbing will be recovered when the top asperity reaches the peak of the underlying asperity and then descends, or:

* Leslie also anticipated the global warming caused by the cumulative effect of solar radiation, though he considered that human activity '… can have no influence whatever in altering the average of temperature [of Earth]' (Leslie, 1804).

† Leslie's arguments were percolating very slowly through the scientific community. According to the *Encyclopædia Britannica* (1823), friction is caused either by climbing over asperities, or the breaking of them.

> Friction is, therefore, commonly explained on the principle of the inclined plane, from the effort required to make the incumbent weight mount over a succession of eminences. But this explication, however currently repeated, is quite insufficient. The mass which is drawn along is not continually ascending; it must alternately rise and fall, for each superficial prominence has a corresponding cavity. … Consequently, though the actuating force might suffer a perpetual diminution in lifting up the weight, it would, the next moment, receive an equal increase by letting it down again. …

3. *Adhesion.* Adhesion between rubbing surfaces is perpendicular to them. However, friction retards tangential motion, hence adhesion is not associated with friction, or:

> Adhesion appears still less capable directly of explaining the source of Friction. A perpendicular force acting on a solid, can evidently have no effect to impede its advance; and though this lateral force, owing to the unavoidable inequalities of contact, must be subject to a certain irregular obliquity, the balance of chances must on the whole have the same tendency to accelerate as to retard the motion.

   Kragelsky and Shchedrov (1956) pointed out that tangential components of adhesion were unknown in Leslie's times.

Leslie proposed an alternative theory whereby friction is caused by the continuous change of surface shape. During sliding, the asperities on a surface deform both themselves and those on the opposing surface, and push forward the debris formed by this deformation: 'Its existence betrays an unceasing mutual change of figure, the opposite planes, during the passage, continually seeking to accommodate themselves to all the minute and accidental varieties of contact. The one surface, being pressed against the other, becomes, as it were, compactly indented, by protruding some points and retracting others.'

This theory had its supporters. Avitzur (1989) investigated it as a notion of *'mobile ridge'*, or *'wave model'*. Black, Kopalinsky, and Oxley (1990) proposed a theory that the frictional force opposing the sliding of a hard metal surface over a softer one, is the force needed to push the waves of plastically deformed material on the soft surface ahead of asperities on the hard one.

Leslie also analysed the role of lubrication: 'The intervention of a coat of oil, soap, or tallow, by readily accommodating itself to the variations of contact, must … lessen the angles, or soften the contour, of the successively emerging prominences, and thus diminish likewise the friction which thence results.' However, he argued that with lubrication, friction increases with speed, whether the lubricant is a liquid or soap or tallow. This is indeed observed in the regime of hydrodynamic lubrication (see Section 4.1).

Interestingly, a quarter century later, Leslie (1829) was less determined in refuting the link between adhesion and friction:

> Most solid bodies, when brought close together, are disposed to cohere mutually, and with various degrees of tenacity. This peculiar force, being exerted perpendicular to the surface of contact, can evidently have no influence whatever in impeding a lateral traction. But all substances appear to possess likewise a certain adhesive property, which opposes any change of mutual contact, and retards even the horizontal passage of one plane along another. This latent obstructing power constitutes Friction, which has such extensive influence in diminishing the performance of all machinery.

Nevertheless, he maintained that the link is by no means certain:

> The obstruction which a loaded carriage has to overcome, when drawn along a smooth level road, is always composed of two very distinct portions; first, the attrition of the axle against the box of the nave, and, secondly, the adhesion of the rim of the wheel as it rolls over the yielding surface of the ground. These elements of retardation, though quite different in their nature, have been often confounded under the general term friction. But it would evidently be rash to infer the properties of adhesion from experiments made on ordinary friction.

3.1.2 George Rennie

Rennie (1791–1866) was a successful engineer and entrepreneur, and wrote an important paper on friction in 1829 (Rennie, 1829). Kragelsky and Shchedrov (1956) contended that he only reported observations without any theoretical analyses, but his experiments and findings are noteworthy. He experimented with more materials than Coulomb (ice, textile, leather, wood, stone and metals) using sophisticated apparatus, and concluded the following:

1. 'The laws which govern the retardation of bodies gliding over each other are as the nature of those bodies. (i.e. determined by the properties of contacting surfaces).'
2. 'That with fibrous substances, such as clothes etc., friction is increased by surface and time, and diminished by pressure and velocity.'
3. 'That with harder substances, such as woods, metals, and stones, and within the limits of abrasion, the amount of friction is as the pressure directly, without regard to surface, time, or velocity.'
4. 'That with dissimilar substances gliding against each other, the measure of friction will be determined by the limit of abrasion of the softer substance', consistent with the modern notion that friction is affected by the shear strength of surfaces (see Section 4.4).

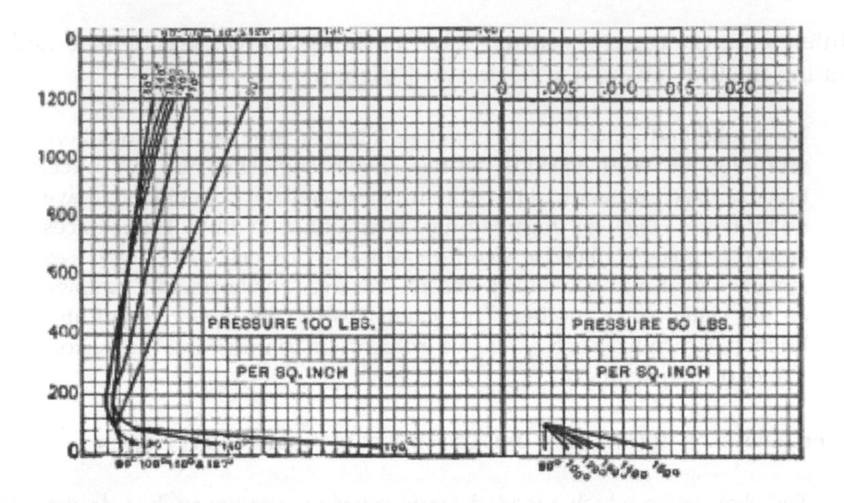

FIGURE 3.1
The dependence of the COF (plotted on the abscissa) on velocity (in feet per minute, plotted along the ordinate) and pressure. (Reprinted from R.H.Thurston, *A Treatise on Friction and Lost Work in Machinery and Millwork*, New York: John Wiley & Sons, 1887. With permission.)

5. 'Friction is greatest with soft, and least with hard substances.'
6. 'The diminution of friction by unguents is as the nature of the unguents, without reference to the substances moving over them.' Basically, this is the case with full film lubrication, where the surfaces are completely separated by lubricant.

3.1.3 Studies of Dry Friction in the Rest of the Nineteenth Century

After Leslie and Rennie, progress in the studies of dry friction was slow. Arthur Morin (1795–1880) was another French military engineer to conduct many careful experiments to determine the COF for various conditions and materials (Dowson, 1998). The obtained coefficients were found useful by practitioners, although Morin did little to investigate the causes of friction. The basic laws of dry friction, quoted in two reputable books from around the turn of the century, were no more informative than the findings of Rennie and Coulomb (Thurston, 1887; Davis, 1904). Probably the most noteworthy development in this period was the systematic study of friction–velocity dependence:

1. Kragelsky and Shchedrov (1956) quoted the experiments with the train braking conducted by Boche in France in 1855 and 1861. It was observed that COF decreases with increasing speed.
2. The same dependence was obtained in the experiments on the English railways in the late 1870s (Galton, 1894).

3. Carefully conducted experiments with wood, stone, steel and pig iron of Conti (Kragelsky and Shchedrov, 1956) in the 1870s showed that the COF first increases with speed, reaches a maximum, then decreases. Conti's two explanations of the peak in the friction dependence of speed are interesting, although not convincing. According to the first one, asperities collide, and the frequency of collisions, hence friction, increases with speed. However, at high speeds the surfaces are polished, and that decreases friction. The second explanation is long-winded and based on the thickness of air film between surfaces.

3.2 Liquid Lubrication

The developments in the studies of liquid lubrication were momentous. They were prompted by the discovery of mineral oil, which promised to be cheaper and more plentiful than the animal and vegetable products (Dowson, 1998). There was also a need for better lubrication of the increasing number of machines, particularly trains. Bearings on rolling stock were performing poorly, and the limitations of lubrication with grease, soap and tallow were frustrating.

The major developments were obviously paving the ground for the Stribeck curve:

1. Gustav Adolph Hirn (1815–1890) studied in 1847 the performance of several lubricants (fats, oils, water and air), and concluded that without lubrication, friction is independent of speed. With lubrication, friction increases with speed and is directly related to the viscosity of the lubricant (Dowson, 1998).
2. Robert Henry Thurston (1839–1903) devised a machine for the testing of lubricants (Thurston, 1887). He noticed that friction initially decreases with pressure, and then increases. The same dependence was clearly obtained for velocity (Figure 3.1).
3. Nikolai Pavlovich Petrov (1836–1920) developed an equation for the calculation of COF between two cylinders separated by a liquid film, in which COF was directly proportional to speed, and inversely proportional to load (Dowson, 1998).
4. Confusion caused by the discrepancies in various studies prompted the Institution of Mechanical Engineers to sponsor a systematic investigation of the liquid lubrication of bearings (Dowson, 1998). A renowned engineer, Beauchamp Tower (1845–1904), was hired and he pointed out that repeatable measurements are obtained only when the bearing is well lubricated.

Hence, inadequate lubrication caused the divergence of results. He obtained the same dependence on speed as did Thurston (Dowson, 1998).

5. Osborne Reynolds (1842–1912) was intrigued by Tower's reports, and postulated that a sufficiently thick oil film can fully separate solid surfaces, in which case friction can be modelled using the laws of hydrodynamics (Reynolds, 1886). The model agreed with the results by Tower, and has been applied in a modified form to many practical purposes.

References

Avitzur, B. 1989. The effect of surface irregularities, subsurface layers, pressure, lubrication, and sliding speed on friction resistance to sliding in metals. *Key Eng. Mat.* 33:1–16.

Black, A.J., E.M. Kopalinsky, and P.L.B. Oxley. 1990. Sliding metallic friction with boundary lubrication: An investigation of a simplified friction theory and of the nature of boundary lubrication. *Wear* 137:161–174.

Davis, W.M. 1904. *Friction and Lubrication. A Hand-Book for Engineers, Mechanics, Superintendents and Managers*. Pittsburgh: Lubrication Publishing.

Dowson, D. 1998. *History of Tribology,* 2nd ed. London and Bury St Edmunds: Professional Engineering Publishing.

Encyclopædia Britannica vol. IX, 6th ed. 1823. Edinburgh: Constable.

Galton, D.S. 1894. *The Effect of Brakes upon Railway Trams*. Pittsburgh: Westinghouse Air Brake.

Kragelsky, I.V. and V.S. Shchedrov. 1956. *Development of the Science of Friction: Dry Friction*. Moscow: Academy of Sciences of USSR.

Leslie, J. 1804. *An Experimental Inquiry into the Nature and Propagation of Heat*. Edinburgh: Mawman.

Leslie, J. 1829. *Elements of Natural Philosophy, Including Mechanics and Hydrostatics,* 2nd ed. London: Oliver and Boyd, and G.B. Whittaker.

Rennie, G. 1829. Experiments on the friction and abrasion of the surfaces of solids. *Phil. Trans.* 119:143–170.

Reynolds, O. 1886. On the theory of lubrication and its application to Mr. Beauchamp Tower's experiments, including an experimental determination of the viscosity of olive oil. *Phil. Trans.* 177:157–234.

Thurston, R.H. 1887. *A Treatise on Friction and Lost Work in Machinery and Millwork.* New York: John Wiley & Sons.

4

Twentieth Century and Beyond

Tribology made huge advances in the twentieth century. It started with the research of liquid lubrication by Stribeck, which, with the contribution of others, resulted in the Stribeck curve. The molecular theory of friction was resurrected. Hardy investigated and baptised the boundary friction, and Bowden and Tabor created an enormous body of work. Their efforts were particularly important in explaining the origin of dry friction via adhesion and the ploughing of asperities. New instruments enabled the investigation of the fundamentals of friction on the microscopic and atomic scales. Finally, substantial progress was made in the mathematical modelling of friction and the development of solid lubricants.

4.1 Stribeck Curve

The development of concepts underlying the Stribeck curve is a fine example of simultaneous and independent research of the same subject by different people. In the first quarter of the twentieth century, quite a few scientists investigated the liquid lubrication in bearings, and their work eventually merged into what is known today as the Stribeck curve (Figure 4.1)*:

1. Richard Stribeck (1861–1950) investigated sliding and rolling bearings, measuring friction as a function of load, speed and temperature. In order to remove the influence of temperature on viscosity, he recalculated the measured COF as a function of load and velocity for a constant bulk-oil temperature of 25°C (Czichos, 1978). The experiments were finished around 1902 (Kragelsky and Shchedrov, 1956).
2. Arnold Sommerfeld (1868–1951), a renowned quantum physicist, solved the equations in the Reynolds model more elegantly, and validated it with Stribeck's data (Czichos, 1978).
3. According to Ciulli (2001), Gümbel arranged Stribeck's data into a curve in 1914. Jones (1985) argued that Stribeck's copious experimental

* Actually, the curve produced by Thurston in 1887 (Figure 3.1) already resembled the Stribeck curve. In 1888, Martens examined the dependency of friction on what is now called the Stribeck number (Woydt and Wäsche, 2010).

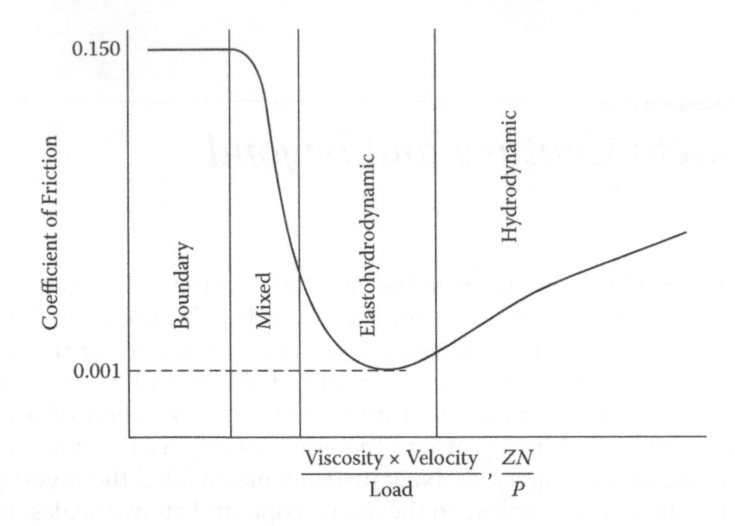

FIGURE 4.1
Distinct regimes of lubrications in the Stribeck curve. (Reprinted from W.R. Jones and M.J. Jansen, NASA TM-209924, 2000.)

data were hard to condense in a useful form. So, Mayo Hersey (1886–1978) conducted similar tests and devised a graph plotted against a dimensionless number.

4. Biel noted in 1920 that Stribeck's data can describe the behaviour of lubricated surfaces if presented as a function of viscosity, sliding velocity and load (Czichos, 1978).
5. There are some potentially confusing issues with the curve, and Appendix B provides additional information.

4.2 Ludwig Gümbel

In a work published posthumously in 1925, Gümbel (1874–1923) proposed his theory, where friction is a sum of dry friction and abrasion (Kragelsky and Shchedrov, 1956). Dry friction dominates if the pressure is low enough to keep the deformation of asperities elastic, and the COF is constant. At the pressure high enough for asperities to deform plastically, abrasion starts. The pressure can be represented as the sum of the elastic component p_1 and the plastic component p_2:

$$p = p_1 + p_2 \tag{4.1}$$

and the resistance to motion is:

$$\tau = f_1 p_1 + f_2 p_2 \tag{4.2}$$

where f_1 and f_2 are the coefficients of dry and abrasive friction, respectively. If the elastic pressure is close to the total pressure, the total COF [Equation (4.3)] is close to the value obtained by Amontons. Kragelsky and Shchedrov praised Gümbel for his effort to develop a practical quantitative model of friction, and credited him with the ideas of cold welding, the protective role of adsorbed gases or liquids, and the link between molecular forces and friction.

$$f = \frac{\tau}{p} = f_2 - (f_2 - f_1)\frac{p_1}{p} \tag{4.3}$$

4.3 Resurrection of the Molecular Theory of Friction

Leslie rejected the theory of adhesion in 1804, using what looked then an unassailable argument. Marcel Brillouin (1854–1948) proved that, theoretically at least, adhesion could be a cause of friction (Brillouin, 1899). He proposed that the adhesion between molecules on the surfaces of rubbing bodies can occur during tangential movement; during sliding, there is a continuous exchange of connections between bodies. Also, these phenomena produce the heat observed during friction. The molecular theory was subsequently revived, particularly by the efforts of Hardy, Tomlinson and Deryagin.

4.3.1 William Bate Hardy and the Concept of Boundary Friction

Hardy (1864–1933) graduated as a zoologist, then studied histology, colloids, then friction and, finally, adhesion. He was also a capable scientific administrator as the chairman of the Food Investigation Board in the Department of Scientific and Industrial Research of the United Kingdom (Hopkins and Smith, 1934). Kragelsky and Shchedrov (1956) and Dowson (1998) regarded his work highly, and praised the meticulous preparation of experiments. Extreme attention was paid to the cleanliness of surfaces and the control of humidity. His key contributions to tribology are the discovery of boundary lubrication and the resurrection of the molecular theory of friction. He reported that very thin films, perhaps 'only one or a very few molecules thick', provide good lubrication (Hardy, 1920). Hardy and Doubleday (1922) elaborated on this phenomenon and christened it, '*What Osborne Reynolds calls "boundary conditions" then operate, and the friction depends not only on the lubricant, but also on the chemical nature of the solid boundaries*'.

Regarding the molecular theory, Hardy (1920) stated: '... the friction both of lubricated and of clean faces is due to true cohesion ... which binds together the molecules of a solid or of a fluid'. Hardy and Doubleday (1922) refuted Coulomb's assertion 'that friction is due to asperities acting like inclined planes', because very smooth surfaces exhibit high friction. Instead, 'The asperities required by Coulomb are in fact the atoms and molecules'. Two other important notions are as follows:

1. Amontons' law was found to hold as long as the properties of surfaces do not change, '... it is a rigid law for hard solids such as glass and hard steel'. However, the law fails with the surfaces that exhibit viscous flow under high pressure, such as wood (Hardy and Doubleday, 1922).
2. Hardy (1920) experimentally established that oxide films provide efficient lubrication on copper. As shown later, the lubricity of oxide is extremely important for the understanding of frictional phenomena in metal rolling.

4.3.2 Tomlinson, Frenkel, Kontorova and Deryagin

In 1929, George Arthur Tomlinson (1885–1943) modelled the friction on the atomic scale in a way popularly explained by Silin (1987). The atoms are likened to caryatides. Initially, shorter caryatides do not touch the load. As the load increases, more caryatides support it, inasmuch as the taller ones are compressed, and their number is proportional to the load, so the macroscopic friction follows Amontons' law. Silin also believed that Tomlinson was the first to explain plausibly why rolling friction is much smaller than sliding friction. It was assumed that both types of friction are caused by the adhesion of surfaces. At sliding, all asperity joints between the surfaces break at the same time. At rolling, only the joints over the contact area are broken.

Tomlinson also outlined the model for the calculation of friction between the same and different materials. However, some of the parameters necessary for the quantification were not known at that time. Nevertheless, the model is considered to be an important step forward. The conceptual idea is simple: the atoms on one surface are represented as particles attached with springs to a substrate, moving over a corrugated surface (Figure 4.2a). It is sometimes called the Prandtl–Tomlinson model, although Popov and Gray (2012) pointed out that Ludwig Prandtl (1875–1953) formulated a more articulate version in 1928.

Soviet physicists Yakov Frenkel (1894–1952) and Tatyana Kontorova (1911–1977) developed a model with atoms, interconnected with springs, bumping over a substrate (Figure 4.2b). The models were combined into the Frenkel–Kontorova–Tomlinson model, where atoms are connected both to the substrate and to the other atoms with springs (Figure 4.3). This model

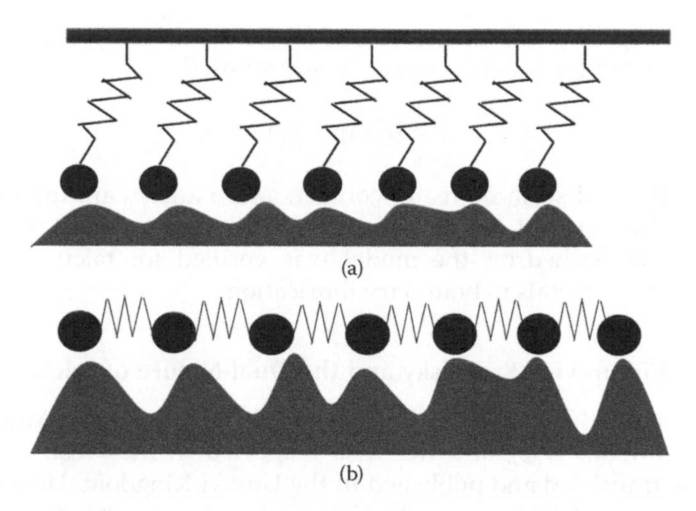

FIGURE 4.2
Models by (a) Tomlinson, (b) Frenkel–Kontorova.

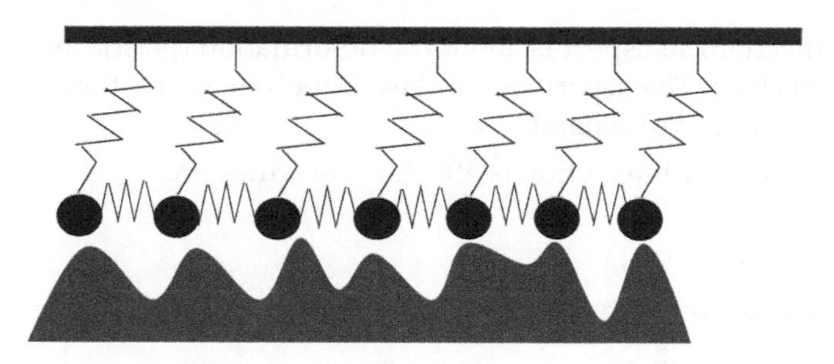

FIGURE 4.3
Frenkel–Kontorova–Tomlinson 2D model.

is widely used; recent modifications were discussed by Robbins (2001). However, Pogorelov (2003) claimed that the Frenkel–Kantorova model cannot be applied to lubrication. Aichele and Mueser (2003) contended that in boundary lubrication atoms are weakly connected to each other and to confining walls, so bonds can be broken. In the elastic models, such as Frenkel–Kontorova–Tomlinson, this breakage is not allowed to occur.

Boris Vladimirovich Deryagin (1902–1994)* extended Tomlinson's model with the interaction between crystalline surfaces described by a statistical

* Deryagin made a huge contribution to the science of colloids and the theory of adhesion (Roldugin, 2006). His working habits were legendary; he was capable of visiting a co-worker at 10 p.m. on 31 December to discuss a problem, and returning at 9 a.m. on 1 January to pester him with preliminary results. Unfortunately, he was involved in the research of 'polywater', where it was claimed that under certain conditions water assumes unusual properties. The results were wrong, presumably due to contaminated samples. Although he recanted his views, this error delayed his admission into the Soviet Academy of Sciences, and probably cost him the Nobel Prize.

approach (Kragelsky and Shchedrov, 1956; Dowson, 1998). The following expression for the total friction force F was proposed:

$$F = \mu S (p + p_0) \quad [4.4]$$

where S is the real surface area of contact, and p and p_0 are the pressures caused by the external forces and adhesion, respectively. According to Kragelsky and Shchedrov, the model was verified for talcum, graphite, quartzite and the metals in boundary lubrication.

4.3.3 Igor Victorovich Kragelsky and the Dual Nature of Friction

Kragelsky* (1908–1989) worked in an institute of the Soviet Academy of Sciences (Dowson, 1998), and was also respected in the West; two of his books were translated and published in the United Kingdom. He advocated the dual nature of friction, mechanical and molecular (Shchedrov, 1949; Kragelsky and Shchedrov, 1956):

1. The mechanical aspect is about the deformation (plastic and elastic) of asperities. The asperities can also penetrate each other, so shearing is required to separate them.
2. The molecular interaction is effected via adhesion.

4.4 Bowden and Tabor

Frank Philip Bowden (1903–1968) and David Tabor (1913–2005), assisted by many able collaborators, produced a large and influential body of work (Appendix C). Kragelsky and Shchedrov (1956) praised Bowden for 'interestingly designed experiments'. Many other researchers were perceived as being simply interested in obtaining COF, without thinking about the causes of friction. However, 'Bowden ... investigated separate processes of which the friction consists, which significantly deepened our understanding of the nature of friction'.

* Little is known about Kragelsky's associate Shchedrov who languished in academic backwaters, and wrote a book on the modelling of elastic fibres. A student remembered him as a quiet genius of unremarkable appearance, drinking heavily, presumably due to romantic problems (Zil'berberg, 2013). Shchedrov lived in his own world of mechanics and mathematics, and his face 'bore a stamp of suffering'. He smiled rarely, but was ecstatic when writing some important theorem on the blackboard; 'his face emitted light, his eyes were burning, his voice was trembling'. While deriving a theorem of Lagrange, he became so excited, that, while writing final equations, he announced to students, 'Now you will witness the birth of a miracle, the miracle of unusual beauty', and was so overwhelmed with excitement that he had to leave the lecturing room and could only continue after several days.

4.4.1 Key Contributions

4.4.1.1 Real Contact Area

The independence of friction force on the area of contact had been a puzzle for centuries. Tabor (1969) pointed out that it had been believed that surfaces touch each other over the whole contact length, like a jigsaw puzzle (Figure 2.5a). Bowden and his team observed that the real area of contact is much smaller. For steel it could be 10,000 times smaller than the apparent area (Bowden, 1952), because surfaces touch each other at the tips of asperities (Figure 2.5b), or as Bowden explained it, 'like turning Switzerland upside down and standing it on Austria—the area of intimate contact will be small' (Tabor, 1969). It was experimentally shown that the real area of contact is:

1. Almost independent of the size of the surfaces, and is very little influenced by the shape and degree of the roughness of the surfaces.
2. Directly proportional to the load. Even at light loads the local pressure at the tiny points of contact is so high that hard metals may flow plastically. The peaks of asperities are crushed until their contact area is big enough to support the applied load.

4.4.1.2 Key Mechanisms of Friction

Bowden and Tabor postulated that the friction force consists of two components, namely the ploughing of hard asperities through the softer opposing surface, and the shearing of the junctions formed by the adhesion between the asperities of the rubbing bodies. The adhesion is caused by high pressure at contact points*.

Mathematically, it can be expressed as follows (Bisson, 1968):

$$\textit{Frictional force} = \textit{shear} + \textit{ploughing} = As + A'p \tag{4.5}$$

where A is the real area of contact, A' ploughing area, s shear strength and p flow pressure. The real area of contact is given by:

$$A = \textit{load}/\textit{flow pressure} = W/p \tag{4.6}$$

and the coefficient of friction (COF) is then:

$$\mu = \frac{\textit{frictional} \quad \textit{force}}{\textit{load}} = \frac{As}{W} + \frac{A'p}{W} \approx \frac{s}{p} \tag{4.7}$$

* The importance of adhesion was supported by the experiments of McFarlane and Tabor (1950a,b) and Rabinowicz and Tabor (1951). When similar metallic surfaces get closer, metallic bonds form. Bonds will form with dissimilar metals too, the strength of the bond being between the bond strengths of the components (Tabor, 1975).

This formula suggests that a low-shear-strength film reduces the COF by reducing the ratio of *s* and *p*. Those films can be formed by oxidation, adsorption of oxygen or moisture, or coating with a softer metal. McFarlane and Tabor (1950a) suggested that oxides and lubricants attenuate adhesion between metals; generally, what reduces adhesion, reduces friction as well. As for the ploughing, it is caused by the displacement of the softer of the two metals by an asperity of hard metal, and is often much smaller than the shear term.

Incidentally, various authors differently define the parameters in these equations:

1. Dowson (1998) defined *s* as shear stress, and *p* as hardness of softer material. For metals, the shear stress is $s = 0.5\sigma_y$ and $p = 3\sigma_y$, where σ_y is the yield stress in tension. The resulting COF for clean metals is ~1/6.
2. Larsen-Basse (1992) defined *s* also as shear stress, and *p* as flow hardness. The flow hardness is about three times greater than flow stress, whereas shear stress is 50–60% of flow stress. Hence, COF for common materials is around 0.17–0.2.

These COFs are of the same order of magnitude as those obtained by early experimenters, but several times smaller than the values obtained for clean metals, where adhesion is much stronger than at contaminated surfaces. Most metals are covered by oxides, adsorbed oxygen or moisture when exposed to air, and so have a similar low COF. Even a monolayer of adsorbed oxygen can hinder adhesion (Tabor, 1975).

4.4.1.3 Soft Films, Oxides and Lubrication

Kragelsky and Shchedrov (1956) credited Bowden and Tabor for the use of films of soft materials to reduce friction (see the anecdote in Appendix C). Such films ensure a smaller surface contact area of rubbing materials and low shear strength, hence low COF according to Equation (4.7). However, with a hard oxide on a soft metal, the metal can break easily under the load, leading to the cracking of the oxide, and exposure of the metal surface. If hardness is similar, both substrate and oxide will deform, the breakthrough of the surface will not occur and COF will stay low. Tabor (1975) also stated that ductile oxide sticks to metal, whereas the brittle one cracks, promoting metal-to-metal contact, hence adhesion.

Figure 4.4 shows that COF decreases with film thickness, due to the reduced direct contact between metals and the subsequent adhesion. In both cases, though, after reaching a minimum at ~0.5 μm, the COF tended to increase with thickness for unspecified reasons. Jones (1985) noted the similarity with the Stribeck curve, without further analysis. Perhaps at the higher thickness, the film has such a low shear strength that it starts to behave as a liquid lubricant, and follows the Stribeck curve.

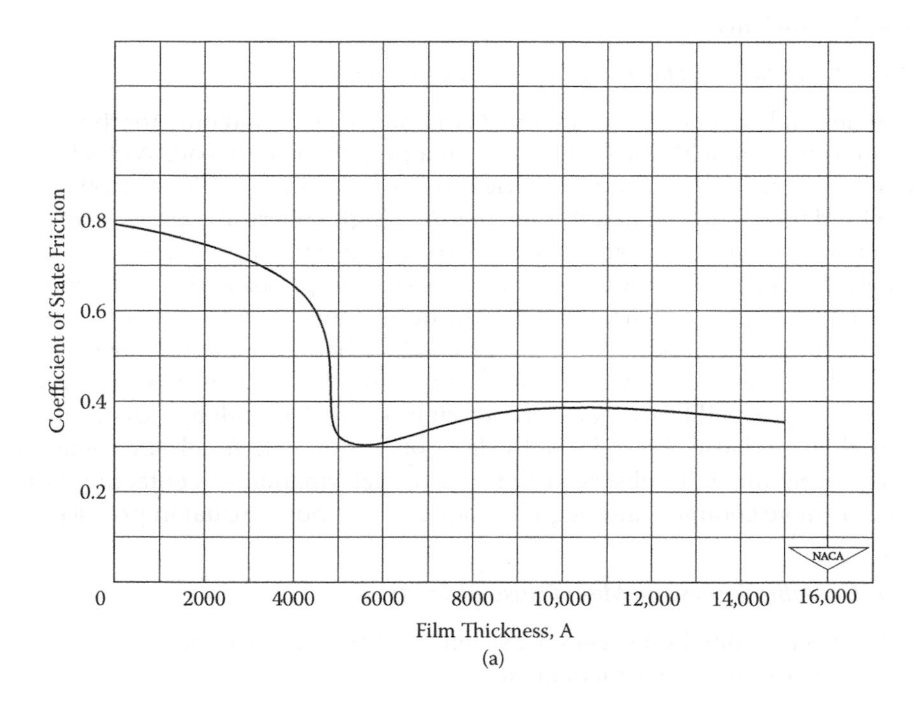

(a)

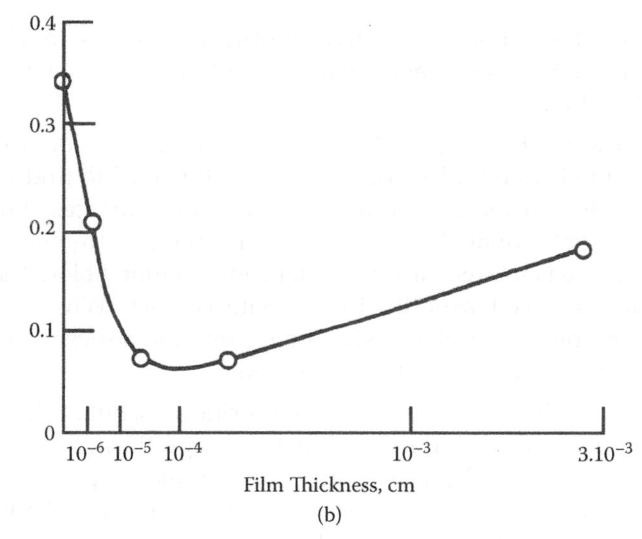

(b)

FIGURE 4.4
COF as a function of thickness of (a) FeS on steel. (Reprinted from E.C. Levine and M.B. Peterson, *Formation of sulfide films on steel and effect of such films on static friction*. NACA TN-2460, 1951.) (b) Indium film on steel. (Adapted from F.P. Bowden and D. Tabor, *The Friction and Lubrication of Solids*, Oxford: Oxford University Press, 1950. Reproduced with permission from Oxford University Press.)

4.4.2 Criticisms

4.4.2.1 Plastic and Elastic Deformation of Asperities

Bowden and Tabor (1939) argued that Amontons' laws hold only for the plastic deformation, with the real contact area proportional to load. With elastic deformation, the contact area varies to the 2/3 power of load. However, Archard (1957) showed that this applies to a single spherical asperity pressed against a flat plate. Where many spherical asperities are in contact, with smaller asperities on the top of larger ones, like pimples, the real contact area becomes proportional to load as the number of asperity layers increases. Furthermore, he reported the experiments where Amontons' laws were obeyed for elastic deformation. Greenwood and Williamson (1966) used the Gaussian probability distribution of height to model randomly distributed asperities, and also showed that the laws of friction hold for elastic contacts. They experimentally observed that plastic deformation of contacting surfaces is more common, although the elastic one is not unusual in practice.

4.4.2.2 Adhesion as the Main Cause of Friction

It has been claimed on several occasions that the theory of adhesion as the main cause of friction is inadequate:

1. Gretz and Bickerman presented lengthy arguments in the comments on the Merchant paper (1968), though they were convincingly refuted by Tabor.
2. Larsen-Basse (1992) argued that adhesion might not be a distinct component of friction. Except in vacuum, it is hard to find measurable adhesion between common engineering surfaces. First, the surface is contaminated, so it is hard to form metal-to-metal bonds. Second, there is a large elastically deformed region below the small plastically stressed asperity in intimate contact. When the load moves, the release of elastic strain ruptures the adhesive bond, so significant adhesion cannot be measured.
3. Rabinowicz (1995) summarised the key criticisms, but outlined their shortcomings. He also contended that although the adhesion theory is essentially plausible, it gives an oversimplified representation of frictional phenomena. It was also noted that other contributions to friction, such as ploughing, roughness and obscure electrical phenomena, are practically negligible.
4. Ludema (1996) described the theory as incomplete, inasmuch as it is not useful for predicting the COF. He conceded that it is superior to the theory of interlocking asperities, and is supported by observed sticking and high COF of clean metal surfaces.

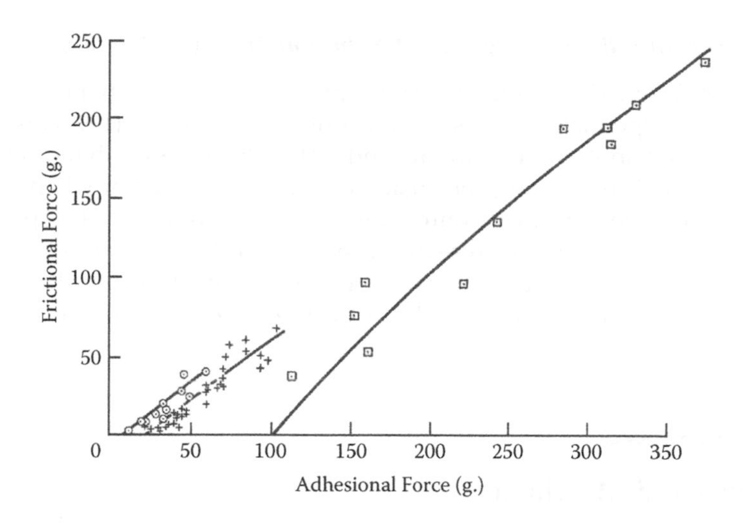

FIGURE 4.5
Friction force versus adhesion force between steel ball and indium. Different curves were obtained with different loads. (Reprinted from J.S. McFarlane and D. Tabor, *Proc. R. Soc. Lond.* A 202:244–253, 1950. With permission from the Royal Society.)

Despite these criticisms, adhesion and ploughing are still viewed as the key causes of macroscopic friction (Chaudhury, 1996; Plößl and Kräuter, 1999; Bonny, De Baets, and Vleugels, 2009). The concept of adhesion was successfully applied to the hot rolling of steel (Section 4.8), and McFarlane and Tabor (1950b) observed a strong correlation between the friction and adhesion forces (Figure 4.5).

4.4.2.3 Criticism by Kragelsky and Shchedrov

The key criticism was that Bowden and Tabor generalised their findings, whereas they are valid only under certain conditions:

1. Occurrence of high temperatures, even localised melting, was assumed at the point of asperity contact, whereas the temperatures that high occur in certain cases only.
2. Plastic deformation of asperities was assumed, although it could often be elastic. This criticism was also voiced by Archard (1957) and Greenwood and Williamson (1966), as discussed above. The latter authors observed that plastic deformation is more common in engineering practice, however.
3. In Equation (4.5), the shear strength was presumed constant, despite its dependence on pressure and its variations over the area of contact. However, that model was intended to be qualitative anyway, thus this would not be a major drawback.

4.4.2.4 Role of Others in Creation of Adhesion Theory of Friction

Bisson (1968) and Komanduri (2006) pointed out that Merchant developed the theory independently and simultaneously. Rabinowicz (1995) also mentioned the contribution of Holm around 1940. On the other hand, Ludema (1996) cautioned that wrong conclusions are easily arrived at when dealing with 'immature ideas: … the conflicting claims are supported by "proof" of prior publication of ideas or research results … full credit should not go to one who does not adequately convince others of his ideas. On this ground alone, Bowden and Tabor are worthy of the honor accorded them.'

4.5 Ernst and Merchant

Hans Ernst (1892–1978)* and Eugene Merchant (1913–2006) investigated friction in 1930s and 1940s, and were highly regarded by Kragelsky and Shchedrov (1956), and Bowden and Tabor (1950). They studied friction between chip and tool in metal cutting; their model and its experimental validation are discussed in Section 9.2. Their views can be summarised as:

1. Only plastic contact occurs, and stress at the contact depends on hardness, not on load.
2. The COF for ideal smooth surfaces is the ratio of shear strength and hardness. In boundary friction, the COF is the tangent of the angle between contacting asperities and the direction of friction force.
3. The stick-slip observed is caused by the localised metal melting.

4.6 More Recent Views on Friction at Macroscopic Level

The proposed causes of macroscopic friction were well presented by Czichos (1978), who divided sliding friction into three stages. Elastic or plastic deformation, or ploughing, occurs in the first stage, and adhesion occurs in the second stage. In the third stage, tangential dislocation shears the

* Ernst was a truly cosmopolitan character (born of German parents in Melbourne, where he graduated, eventually moving to the United States), and an engineer for all seasons (he taught in a technical school in Victoria, started in a bicycle repair shop in California, and later held senior positions in industry, research, and academia; Merchant, 2003).

joints formed by adhesion, with possible elastic recovery for elastic deformation. There are several possible mechanisms of adhesion: long-range van der Waals force, which acts between different types of materials, and short-range forces (metallic, covalent or ionic). Clean pieces of the same metal form metallic bonds at contacts, with the interface strength equal to the strength of bulk metal.

Larsen-Basse (1992) proposed several mechanisms of friction, which act simultaneously, or several at a time, with a particular mechanism dominant in certain conditions:

1. Adhesion, which is the dominant mechanism when surfaces are very clean. Cold welding occurs due to the interatomic forces, at very low load, and COF is very high.
2. Plastic deformation and ploughing caused by deformation of the softer surface by the asperities of the harder one.
3. Elastic deformation of material below the plastically deformed regions.
4. Deformation or fracture of surface layers such as oxides. This basically enhances the adhesion of clean surfaces when the protective film is removed.
5. Interference and local plastic deformation caused by third bodies, mainly agglomerated wear particles, trapped between the moving surfaces. These particles may indent surfaces, although, as pointed out by Schey (1983), the friable ones may break into small pieces acting as ball-bearings, and reduce the COF.

4.7 Studies of Friction at Microscopic and Atomic Levels

Investigation at the atomic level was made possible by the development of instruments including surface force apparatus, quartz crystal microbalance, and the lateral force, atomic force and friction force microscopes (Krim 2002; Burke 2003; Tambe and Bhushan, 2005; Hölscher, Schirmeisen, and Schwarz, 2008).

The transition from the macroscopic to smaller scales was neatly illustrated by Hölscher et al. (2008), who graphically presented the difference between the apparent, the true and the nanoscale single asperity contacts (Figure 4.6). Most macroscopic and microscopic tribological phenomena are dominated by the influence of wear, plastic deformation, lubrication, surface roughness and surface asperities, hence 'macroscopic friction experiments are therefore difficult to analyse in terms of a universal theory'. They suggested that for

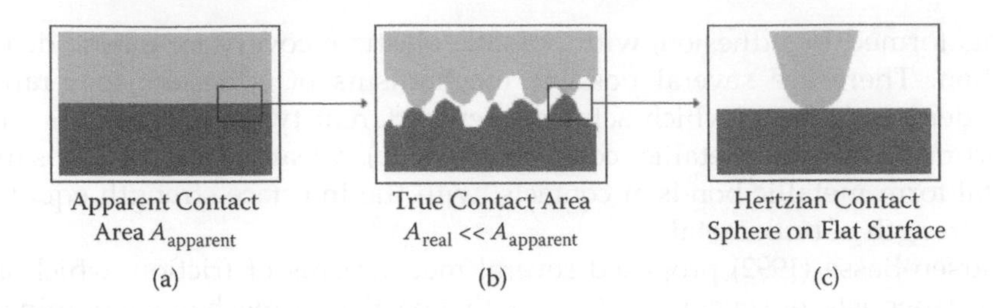

FIGURE 4.6
The difference between (a) apparent contact area observed on macroscopic scale; (b) true contact area with contacts formed between individual small asperities; and (c) nanoscale single asperity contact. (Reprinted from H. Hölscher, A. Schirmeisen, and U.D. Schwarz, *Phil. Trans. A* 366:1383–1404, 2008. Reprinted with permission from the Royal Society.)

the better understanding of friction at the macroscopic level, the frictional behaviour of a single asperity contact should be investigated first. The information obtained that way can then be used to quantify the macroscopic friction statistically, that is, '... by the summation of the interactions of a large number of small individual contacts, which form the macroscopic roughness of the contact interface'.

The macroscopic theory of friction is based on the continuing tearing of junctions formed by adhesion. However, Krim (2002) noted that friction without wear was observed on the atomic scale. She explained this by Tabor's notion, where the atoms close to one surface are set into motion by the sliding atoms on the opposing surface. This generates vibrations (or phonons) that dissipate this energy as sound and heat (Figure 4.7).

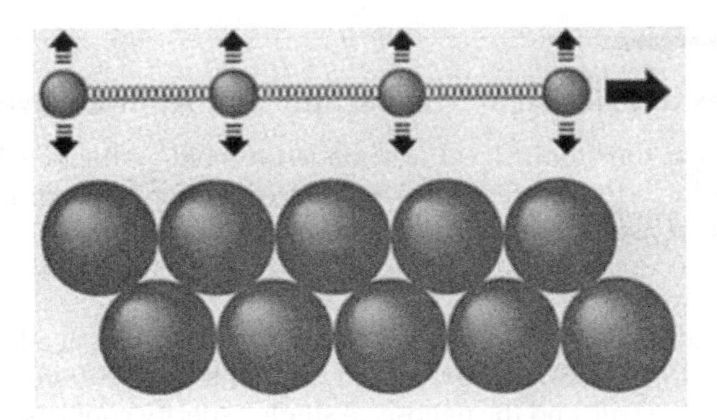

FIGURE 4.7
A single layer of atoms vibrates as it slides over the surface underneath. (Reprinted from J. Krim, *Surf. Sci.* 500:741–758, 2002. With permission from Elsevier.).

4.8 Application of Adhesion Concept to Hot Rolling

Straffelini (2001) contended that the supporters of the adhesion theory usually base COF on the shear strength of the contacting junctions, and this strength is vaguely described in the literature. That motivated an approach where the shear strength is related to the adhesion force between surfaces, and this force is calculated using the thermodynamic work of adhesion. In this way, the COF can be estimated (see Section 9.2). Straffelini also pointed out that contaminants tend to reduce the work of adhesion of metals, thereby reducing friction. The model was validated with experimental data for various metal pairs. The agreement is good (Figure 4.8a), except for cobalt sliding on itself*.

Jupp, Talamantes-Silva, and Beynon (2004) recognised the limitation of this model; it was applied to pure metals, and oxide is always present in the roll gap in hot rolling. They contended that the interface energy is not readily available, and offered a simple model to calculate this term at the steel–magnetite interface. The improved model was incorporated into a finite-element simulation package, which enabled the calculation of COF along the roll gap. Jupp and Beynon (2005) then tested the model experimentally. Although the temperature was well below the one experienced in hot rolling, it was high enough for some oxide to be formed (450–500°C), presumed to be mainly magnetite. The calculated friction for the iron–magnetite interface matched the measured one well (Figure 4.8b).

* Straffelini explained the mismatch with the hexagonal close-packed (hcp) structure of Co, claiming that friction is lower for such metals. This explanation omitted the impact of temperature. The structure of Co changes from hcp to full-centre-cubic (fcc) at 417°C, and fcc Co has high friction (Larsen-Basse, 1992). Thalium similarly undergoes phase transformation from hcp to fcc, and has similar frictional behaviour. However, Larsen-Basse warned that titanium has both hcp structure and high friction.

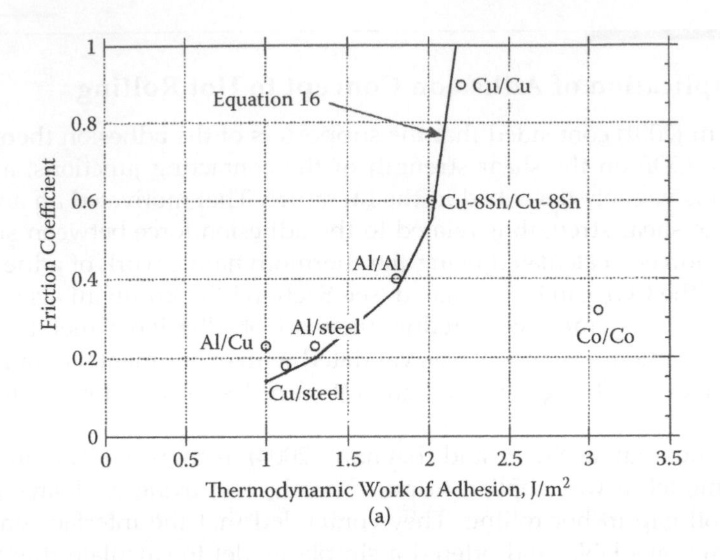

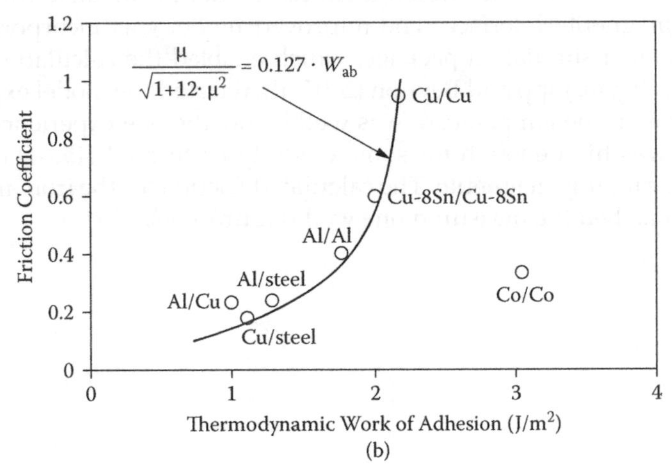

FIGURE 4.8
Experimental validation of Straffelini's model (a) (Reprinted from G. Straffelini, *Wear* 249:79–85, 2001. With permission from Elsevier.); and (b) (Reprinted from S.P. Jupp and J. Beynon, *Steel Res. Int.* 76:387–391, 2005. With permission from Wiley-VCH Verlag GmbH.)

References

Aichele, M., and M.H. Mueser. 2003.Kinetic friction and atomistic instabilities in boundary-lubricated systems. *Phys. Rev.* E68:016125:1–14.

Archard, J.F. 1957. Elastic deformation and the laws of friction. *Proc. R. Soc. Lond. A* 243:190–205.

Bisson, E.E. 1968. *Friction, Wear and the Influence of Surfaces*. NASA TM/X-52380.

Bonny, K., P. De Baets, J. Vleugels et al. 2009. Impact of Cr_3C_2/VC addition on the dry sliding friction and wear response of WC–Co cemented carbides. *Wear* 267:1642–1652.

Bowden, F.P. 1952. Introduction to the discussion: The mechanism of friction. *Proc. R. Soc. Lond. A* 212:440–449.

Bowden, F.P., and D. Tabor. 1939. The area of contact between sliding and between moving surfaces. *Proc. R. Soc. Lond. A* 169:391–413.

Bowden, F.P., and D. Tabor. 1950. *The Friction and Lubrication of Solids*. Oxford: Oxford University Press.

Brillouin, M. 1899. Théorie moléculaire du frottement des solides polis. *Annales de Chimie et de Physique, Septième Série*. XVI:433–456.

Burke, S.A. 2003. *Friction Force Microscopy: Seeking New Understanding of Friction from a Nanoscale Perspective*. Unpublished report. McGill University, 28 February.

Chaudhury, M.K. 1996. Interfacial interaction between low-energy surfaces. *Mat. Sci. Eng*. R16:97–159.

Ciulli, E. 2001. Friction in lubricated contacts: From macro- to microscale effects. In *Fundamentals of Tribology and Bridging the Gap Between the Macro and Micro/Nanoscales*, B. Bhushan (Ed.), 725–734. Dordrecht: Kluwer Academic.

Czichos, H. 1978. *Tribology—A System Approach to the Science and Technology of Friction, Lubrication and Wear*. Amsterdam: Elsevier.

Dowson, D. 1998. *History of Tribology*, 2nd ed. London and Bury St Edmunds: Professional Engineering.

Greenwood J.A., and J.B.P. Williamson. 1966. Contact of nominally flat surfaces. *Proc. R. Soc. Lond. A* 295:300–319.

Hardy, W.B. 1920. Problems of lubrication. In *Notices of the Proceedings at the Meetings of the Members of the Royal Institute of Great Britain with Abstracts of the Discourses Delivered at the Evening Meetings*, 1924: 65–72. London: William Clowes and Sons.

Hardy, W.B., and I. Doubleday. 1922. Boundary lubrication. The paraffin series. *Proc. R. Soc. Lond. A* 100:550–574.

Hölscher, H., A. Schirmeisen, and U.D. Schwarz. 2008. Principles of atomic friction: From sticking atoms to superlubric sliding. *Phil. Trans. A* 366:1383–1404.

Hopkins, F.G., and F.E. Smith. 1934. William Bate Hardy. 1864–1933. *Obit. Not. Fell. R. Soc.* 1:326–333.

Jones, W.R., Jr. 1985. Boundary lubrication–Revisited. In *Tribology: The Story of Lubrication and Wear*, 23–53. NASA TM-101430.

Jones, W.R., and M.J. Jansen. 2000. *Space Tribology*. NASA TM-209924.

Jupp, S.P., and J. Beynon. 2005. A study of friction in non-lubricated high temperature steel processing. *Steel Res. Int.* 76:387–391.

Jupp, S.P., J. Talamantes-Silva, and J.H. Beynon. 2004. Application of fundamental friction model to the hot rolling of steel. In *Metal Forming 2004*, J. Kusiak, P. Hartley, J. Majta et al. (Eds.), 325–329. Bad Harzburg. GRIPS Media.

Komanduri, R. 2006. In Memoriam: M. Eugene Merchant. *Trans. ASME* 128:1034–1036.

Kragelsky, I.V. and V.S. Shchedrov. 1956. *Development of the Science of Friction: Dry Friction*. Moscow: Academy of Sciences of USSR.

Krim, J. 2002. Surface science and the atomic-scale origins of friction: What once was old is new again. *Surf. Sci.* 500:741–758.

Larsen-Basse, J. 1992. Basic theory of solid friction. In *Friction, Lubrication, and Wear Technology, ASM Handbook 18*, 27–38. Ohio: American Society for Metals.

Levine, E.C., and M.B. Peterson. 1951. *Formation of Sulfide Films on Steel and Effect of Such Films on Static Friction*. NACA TN-2460.

Ludema, K. 1996. *Friction, Wear, Lubrication*. Boca Raton: CRC Press.

McFarlane, J.S., and D. Tabor. 1950a. Adhesion of solids and the effect of surface films. *Proc. R. Soc. Lond. A* 202:224–243.

McFarlane, J.S., and D. Tabor. 1950b. Relation between friction and adhesion. *Proc. R. Soc. Lond. A* 202:244–253.

Merchant, M.E. 1968. Friction and adhesion. In *Interdisciplinary Approach to Friction and Wear*, P.M. Ku (Ed.), 181–265. NASA SP-181.

Merchant, M.E. 2003. *An Interpretative Review of 20th Century US Machining and Grinding Research*. Cincinnati: TechSolve.

Plößl, A., and G. Kräuter. 1999. Wafer direct bonding: Tailoring adhesion between brittle materials. *Mat. Sci. Eng.* R25:1–88.

Pogorelov, Y. 2003. Adiabatic theory of boundary friction and stick-slip processes. *arXiv:cond-mat*/0309117:1–8.

Popov, V.L., and J.A.T. Gray. 2012. Prandtl-Tomlinson model: History and applications in friction, plasticity and nanotechnologies. *Z. Angew. Math. Mech.* 92:683–708.

Rabinowicz, E. 1995. *Friction and Wear of Materials*, 2nd ed. New York: Wiley-Interscience.

Rabinowicz, E., and D. Tabor. 1951. Metallic transfer between sliding metals: An autoradiographic study. *Proc. R. Soc. Lond. A* 208:455–475.

Robbins, M.O. 2001. Jamming, friction and unsteady rheology. In *Jamming and Rheology: Constrained Dynamics in Microscopic and Macroscopic Scales*, A.J. Liu, and S.R. Nagel (Eds.), 50–64. London: Taylor and Francis.

Roldugin, V.I. 2006. Boris Vladimirovich Deryagin. *Russian Journal of Chemistry* L(5):134-137.

Schey, J.A. 1983. *Tribology in Metalworking. Friction, Lubrication and Wear*. Ohio: American Society for Metals.

Shchedrov, V.S. 1949. Dry friction and its role in engineering. *Science and Life* 5:22-24.

Silin, A.A. 1987. *Friction and Us*. Moscow: Nauka.

Straffelini, G. 2001. A simplified approach to the adhesive theory of friction. *Wear* 249:79–85.

Tabor, D. 1969. Frank Philip Bowden. 1903–1968. *Biogr. Mems Fell. R. Soc.* 15:1–38.

Tabor, D. 1975. Interaction between surfaces: Adhesion and friction. In *Surface Physics of Materials, Vol. II*, J.M. Blakely (Ed.), 475–529. New York: Academic Press.

Tambe N.S., and B. Bhushan. 2005. Friction model for the velocity dependence of nanoscale friction. *Nanotechnology*, 16:2309–2324.

Thurston, R.H. 1887. *A Treatise on Friction and Lost Work in Machinery and Millwork*. New York: John Wiley & Sons.

Woydt, M., and R. Wäsche. 2010. The history of the Stribeck curve and ball bearing steels: The role of Adolf Martens. Wear 268:1542–1546.

Zil'berberg, I.I. To the Memory of Mark Morozov. http://lit.lib.ru/z/zilxberberg_i_i/zil05.shtml (accessed April 21, 2013).

Section II

Phenomena Relevant to Friction and the Rolling of Hot Metals

5

Roughness and Friction

> … [I]n basic science if an interesting phenomenon is discovered, a serious effort will be made by other research groups to reproduce these results. In many situations in tribology, it is not necessarily that the scientists or engineers are not competent or do not know about controls but there are few research groups, and tribology is dominated by making some piece of mechanical equipment work adequately. This equipment domination often involves very specific applications, with ill-defined conditions, which make repetition of the experiment and comparisons from one laboratory to another very difficult. In fact, when round-tables have been held where attempts have been made to control conditions, the results, with respect to reproducibility of wear, turned out to be poor. The reason may be that the critical parameters for friction and wear have not been determined.
>
> **Ferrante (1987)**

> All things and everything whatsoever however thin it be which is interposed in the middle between objects that rub together lighten the difficulty of this friction.
>
> **Leonardo da Vinci (Dowson, 1998)**

> Unanimity of opinion may be fitting for a rigid church, for the frightened or greedy victims of some (ancient, or modern) myth, or for the weak and willing followers of some tyrant. Variety of opinion is necessary for objective knowledge.
>
> **Feyerabend (1996)**

According to some, roughness helps in reducing friction. Fusaro (1991) argues that most solid films do not bond well to smooth surfaces, and a rough surface serves as a reservoir for lubricant. On the other hand, in some laboratory-scale hot rolling experiments, a strong positive correlation between the COF and roughness was observed (Park, Lee, and Lee, 1999; Kang et al., 2001).

Looking for evidence of a link between friction and roughness, first we discuss the experiments with emery paper and metal. Porgess and Wilman (1959) experimented with emery papers of varying roughness sliding over each other (Figure 5.1). Roughness was expressed as a mean particle diameter; larger diameter meant larger roughness. Avient, Goddard, and Wilman (1960) investigated the friction between emery paper with varying particle sizes and various metals (Figure 5.2). The COF increased with particle size,

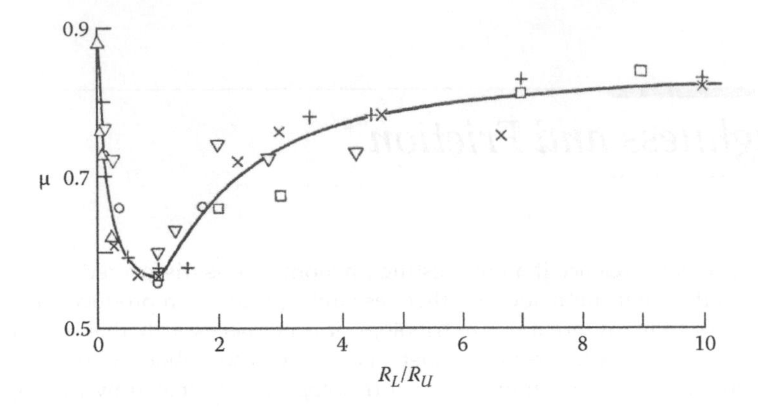

FIGURE 5.1
Friction as a function of the ratio of the particle radii of lower (R_L) and upper (R_U) emery papers. (Reprinted from P.V.K. Porgess and H. Wilman, *Proc. R. Soc. Lond. A* 252:35–44. 1959. With permission from the Royal Society.)

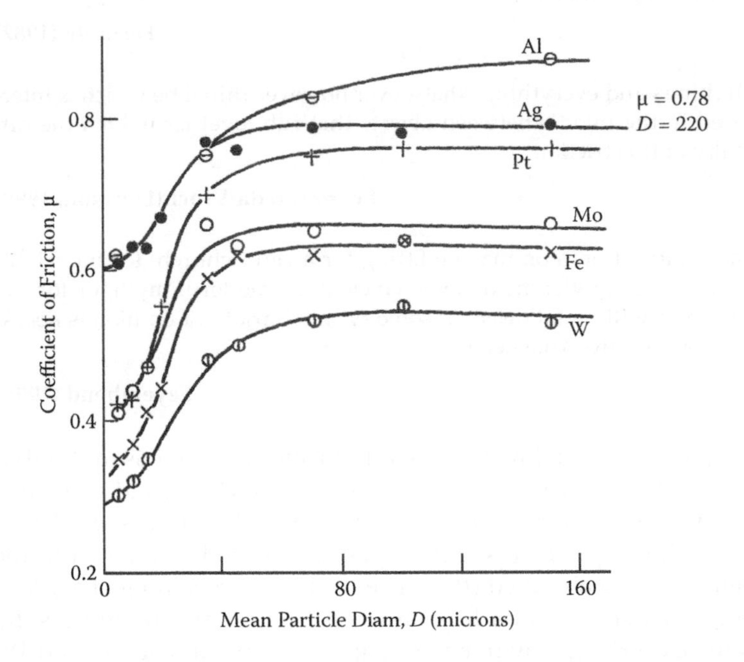

FIGURE 5.2
Sliding of emery paper over various metallic surfaces: the dependence of COF on the mean particle diameter of emery paper. (Reprinted from B.W.E. Avient, J. Goddard, and H. Wilman, *Proc. R. Soc. Lond. A* 258:159–180, 1960. Reprinted with permission from Royal Society.)

as in the second part of the curve in Figure 5.1, eventually commencing in a slight decline for most metals considered.

Both curves suggest that COF increases with roughness; high COF in Figure 5.1 occurs in the cases when one surface is much rougher than the other. On the other hand, Sedlaček, Podgornik, and Vižintin (2009) investigated the friction of an alumina pin on a steel disc. The COF decreased with roughness in dry tests, but it was the opposite case with lubricated surfaces.

Relationship between the COF and roughness was specifically addressed in the investigations of hot strip rolling on laboratory-scale rigs, but the findings were inconclusive. Park et al. (1999) and Kang et al. (2001) established a clear positive correlation (Figure 5.3). Similarly, Azushima, Nakata, and Toriumi (2010) observed that COF increased slightly when roughness increased from Ra = 0.05 μm to Ra = 0.8 μm. Malbrancke, Uijtdebroeks, and Walmag (2007) also suggested a positive correlation, although on a small data sample. On the other hand, Gotoh et al. (1998) observed a negative correlation (Figure 5.4).

These laboratory tests indicate the lack of a clear relationship between roughness and friction in hot rolling. Unfortunately, it was not possible to find any relevant analysis based on plant data in the literature. The observations in a commercial mill reported in Chapter 16 suggest that the chemistry of roll shells is more important for friction than roughness. That is, given the almost identical grinding conditions and similar hardness, roughness cannot explain the large friction differences between some HiCr rolls. Another example is given by Sun et al. (2004), where steel oxidised at 800°C over 80 s has a substantially higher roughness than clean steel. The oxide consists mainly of lubricating oxides FeO and Fe_3O_4, and as shown later they have lower friction than steel. Some general statements in the literature also indicate that roughness would not play a significant role in determining friction in steel rolling, and in general:

1. According to Bowden and Tabor (1950): 'Over wide range of surface finish, the friction of metals is nearly independent of the degree of surface roughness'.
2. Forrester (1946) asserted that the friction of unlubricated surfaces is not dependent on surface finish.
3. Straffelini (2001) contended that the initial roughness has little impact on friction if the load and sliding velocity are high enough to promote rapid polishing.

The divergence between the observations could be plausibly explained by the condition of the surfaces. Although oxidation would not be prominent on emery paper, it could have played a decisive role in the tests conducted with metals in the presence of air.

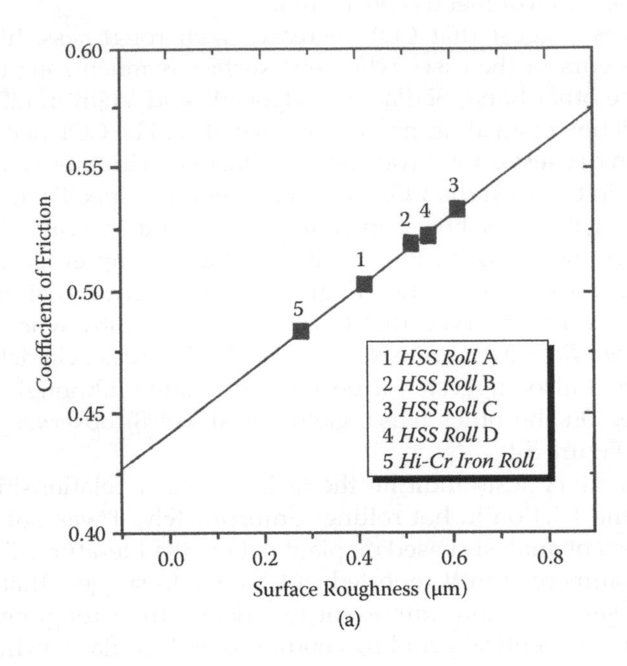

(a)

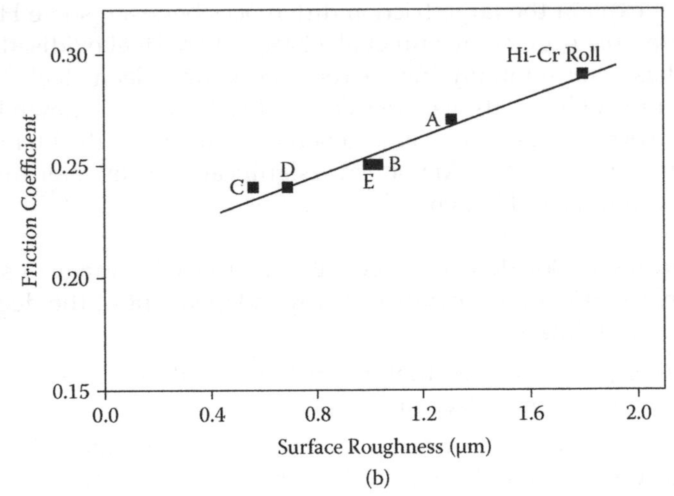

(b)

FIGURE 5.3
Correlation between COF and roughness for high-speed steel (HSS) and high-chromium (HiCr) rolls. ((a) Reprinted from J.W. Park, J.C. Lee, and S. Lee, *Met. Mat. Trans. A* 30A:399-409, 1999; and (b) reprinted from Y.J. Kang, J.C. Oh, H.C. Lee et al. *Met. Mat. Trans. A* 32A:2515–2525, 2001. With permission from Springer Verlag.)

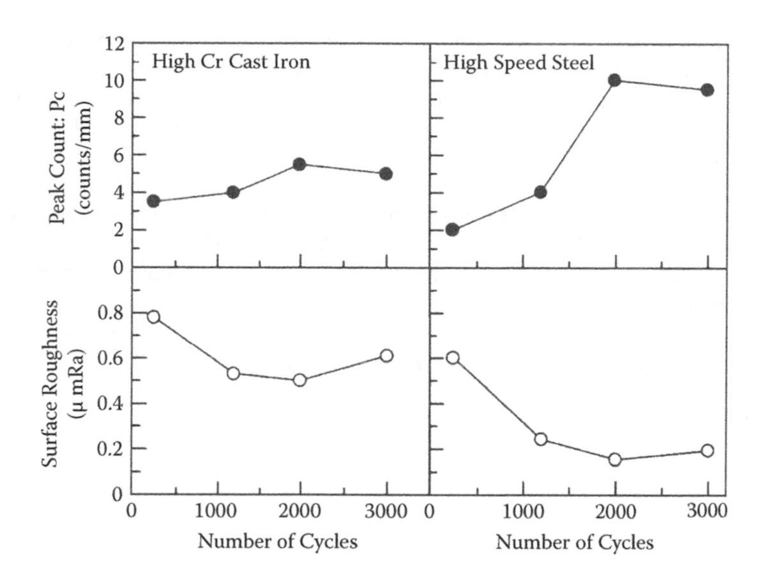

FIGURE 5.4
Change of peak count (closely related to COF) and roughness with a number of disc revolutions on a twin-disc rig. (Reprinted from K. Gotoh, H. Okada, T. Sasaki et al. *Tetsu-to-Hagane* 84:861–867, 1998. With permission from the Iron and Steel Society of Japan.)

References

Avient, B.W.E., J. Goddard, and H. Wilman. 1960. An experimental study of friction and wear during abrasion of metals. *Proc. R. Soc. Lond. A* 258:159–180.

Azushima, A., Y. Nakata, and T. Toriumi. 2010. Prediction of effect of rolling speed on coefficient of friction in hot sheet rolling of steel using sliding rolling tribo-simulator. *J. Mat. Proc. Tech.* 210:110–115.

Bowden, F.P., and D. Tabor. 1950. *The Friction and Lubrication of Solids*. Oxford: Oxford University Press.

Dowson, D. 1998. *History of Tribology*, 2nd ed. London and Bury St Edmunds: Professional Engineering.

Ferrante, J. 1987. *Tribology Theory Versus Experiment*. NASA TM 100198.

Feyerabend, P. 1996. *Against Method*. London, New York: Verso.

Forrester, P.G. 1946. Kinetic friction in or near the boundary region. II. The influence of sliding velocity and other variables on kinetic friction in or near the boundary region. *Proc. R. Soc. Lond.* A 187:439–463.

Fusaro, R.L. 1991. *Tribology Needs for Future Space and Aeronautical Systems*. NASA TM 104525.

Gotoh, K., H. Okada, T. Sasaki, et al. 1998. Effects of roll surface deteriorations on scale defect in hot rolling. *Tetsu-to-Hagane* 84:861–867.

Kang, Y.J., J.C. Oh, H.C. Lee, et al. 2001. Effects of carbon and chromium additions on the wear resistance and surface roughness of cast high-speed steel rolls. *Met. Mat. Trans. A* 32A:2515–2525.

Malbrancke, J., H. Uijtdebroeks, and G. Walmag. 2007. A new breakthrough method for the evaluation of hot rolling work roll grades. *Rev. Met.-CIT* 104:512–521.

Park, J.W., J.C. Lee, and S. Lee. 1999. Composition, microstructure, hardness and wear properties of high-speed steel rolls. *Met. Mat. Trans. A* 30A:399–409.

Porgess, P.V.K., and H. Wilman. 1959. The dependence of friction on surface roughness. *Proc. R. Soc. Lond. A* 252:35–44.

Sedlaček, M., B. Podgornik, and J. Vižintin. 2009. Influence of surface preparation on roughness parameters, friction and wear. *Wear* 266:482–487.

Straffelini, G. 2001. A simplified approach to the adhesive theory of friction. *Wear* 249:79–85.

Sun, W., A.K. Tieu, Z. Jiang, et al. 2004. Oxide scales growth of low-carbon steel at high temperatures. *J. Mat. Proc. Tech.* 155–156:1300–1306.

6

Liquid Lubrication, Stribeck Curve and Friction–Velocity Dependence

Liquid lubrication is introduced before solid lubrication, inasmuch as its interpretation is easier, mainly because of the Stribeck curve (Figure 4.1). The curve shows four lubrication regimes:

1. *Boundary lubrication,* where only a thin layer of lubricant is present, and the load is carried predominantly by asperities.
2. *Partial elastohydrodynamic,* or *mixed, lubrication,* with asperity contact in some areas, and lubricant separation elsewhere. Both the lubricant and the asperities carry the load.
3. *Elastohydrodynamic lubrication,* where the fluid film is present, but if load is high, the lubricant pressure causes elastic deformation of surfaces (Jones, 1985).
4. *Hydrodynamic lubrication,* where surfaces are separated by a continuous film, much thicker than their composite roughness. Friction is caused by the viscous dissipation within the lubricant (Larsen-Basse, 1992), and increases with velocity due to the viscous drag (Ludema, 1996). Wear is much reduced due to the lack of asperity contact, although there is some due to the surface fatigue, cavitation, fluid erosion and the like.

A practical differentiation between regimes is based on the ratio (L) of film thickness to the composite roughness of surfaces. Hydrodynamic lubrication occurs with $L > 3$, whereas boundary lubrication prevails for $L < 1.5$ (Larsen-Basse, 1992).

An interesting explanation of those regimes is given by Holinski (1983) in Figure 6.1, via an analogy with waterskiing. A heavy skier does not float on the top of water, and bumps along the rocky bottom. At higher velocity or lower load, the skier lifts off and has a smoother ride.

It was also argued that roughness amplifies the boundary and mixed lubrication regimes. In Figure 6.2, cases of a rough surface and a surface smoothed with a solid coating are compared. A mixed regime is achieved at higher loads and lower speeds when the surface is coated. A practical example of a taper roller bearing is shown in Figure 6.3.

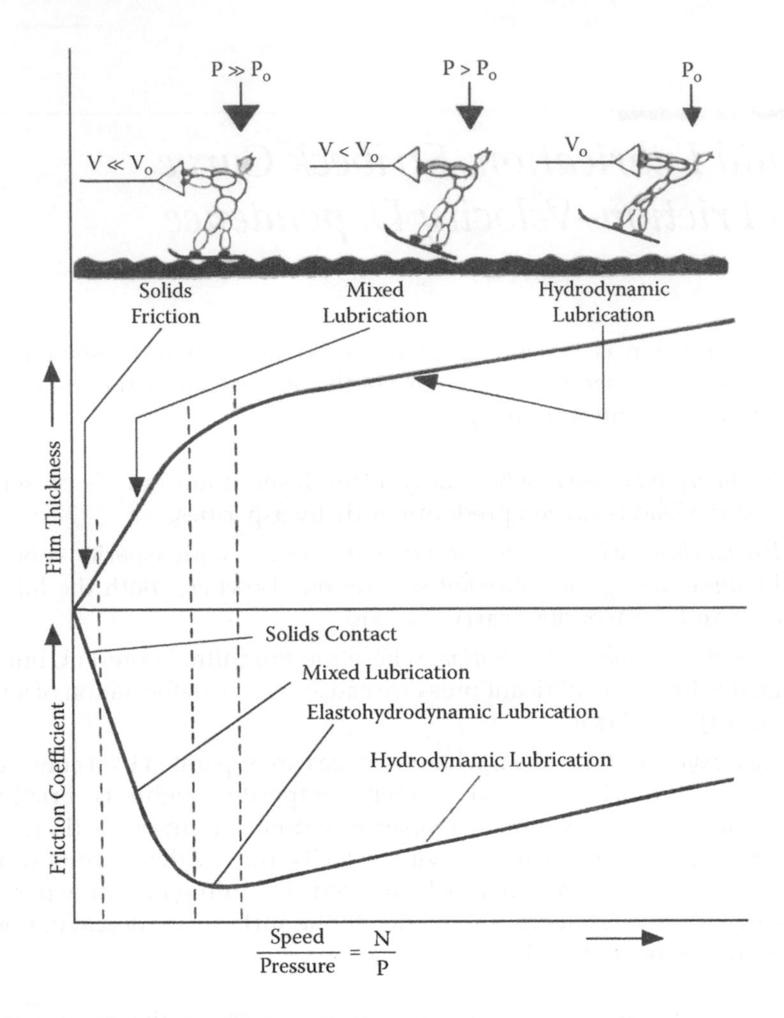

FIGURE 6.1
Interpretation of the Stribeck curve. (Reprinted from R. Holinski, Support of oil lubrication on bonded coatings. In *Tribology in the 80's Vol. 2*, 709–721, 1983, NASA CP-2300.)

The analyses by Holinski explain the impact of load and velocity, but how about viscosity? When the mechanical conditions of velocity and load are kept constant, COF decreases in the hydrodynamic regime as viscosity decreases. However, at low viscosity the capacity to form viscous oil film decreases, and the curve enters the region of the high-friction boundary and mixed lubrication*.

* Persson (1999) pointed out that oil is a better lubricant than water, despite being more viscous and harder to shear. This is because it is easier to squeeze the water out of the contact area, particularly at lower velocities.

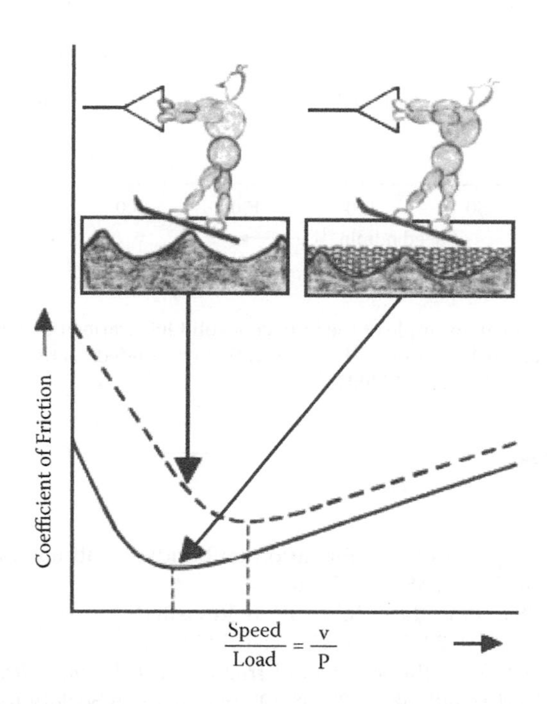

FIGURE 6.2
The Stribeck curve as a function of roughness. (Reprinted from R. Holinski, Support of oil lubrication on bonded coatings. In *Tribology in the 80's Vol. 2*, 709–721, 1983, NASA CP-2300.)

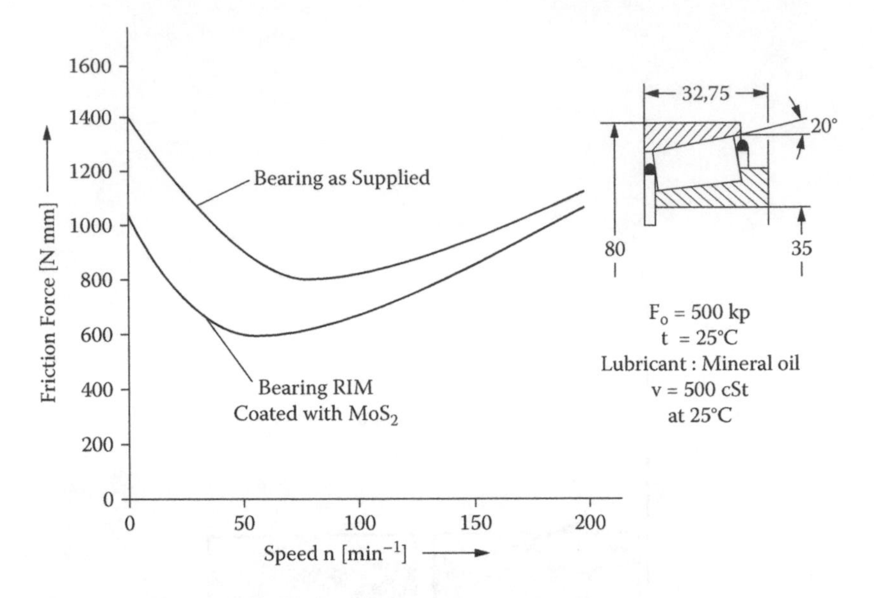

FIGURE 6.3
The Stribeck curve, and an example of the impact of solid lubrication of a taper roller bearing. (Reprinted from R. Holinski, Support of oil lubrication on bonded coatings. In *Tribology in the 80's Vol. 2*, 709–721, 1983, NASA CP-2300.)

References

Holinski, R. 1983. Support of oil lubrication on bonded coatings. In *Tribology in the 80's, Vol. 2*, 709–721. NASA CP-2300.

Jones, W.R. Jr. 1985. Boundary lubrication–Revisited. In *Tribology: The Story of Lubrication and Wear*, 23–53. NASA TM-101430.

Larsen-Basse, J. 1992. Basic theory of solid friction. In *Friction, Lubrication, and Wear Technology, ASM Handbook 18*, 27–38. Ohio: American Society for Metals.

Ludema, K. 1996. *Friction, Wear, Lubrication*. Boca Raton, FL: CRC Press.

Persson, B.N. 1999. Sliding friction. *Surf. Sci. Rep.* 33:83–119.

7

Solid Lubricants

Without liquid lubrication, the only significant lubrication in hot rolling of steel would be by ferrous oxides. It is therefore useful to address the fundamentals of solid lubrication, particularly the dependence of friction on the sliding velocity for various solids.

A key feature of solid lubricants is their performance at high temperatures. Twenty-five years ago, DellaCorte (1987) could not find liquid lubricants to operate above 400°C, and, for space applications, the range of interest was up to 1000°C. He characterized the solid lubricant as a solid that shears easily, providing low friction, yet separating sliding surfaces. Many solid lubricants are laminar, such as MoS_2 or graphite. One sheet sticks to the substrate, and the others slide, having low shear strength (Johnson, Godfrey, and Bisson, 1948). Sliney (1987) believed that a good solid lubricant should be thermodynamically stable, possess a high degree of plasticity, adhere to substrate and have a compatible thermal expansion with it. Hironaka (1984) discussed reaction-generated coating, which is formed by chemical reactions in advance, or during the sliding by frictional heat (*'in-situ* films').

7.1 Impact of Sliding Velocity and Load

Can the Stribeck curve be applied to solid lubrication? Many control engineering models include the Stribeck curve, and are applied to dry friction (Chapter 8). Of three parameters (load, viscosity and velocity), the well-regarded LuGre model explicitly uses only the velocity (Thomsen, 1999; Åström and Canudas-de-Wit, 2008), and no hydrodynamic considerations were used in its formulation (Canudas-de-Wit et al., 1995). In the experimental validation of LuGre, liquid lubrication is not mentioned (Kelly, Llamas, and Campa, 2000; Ferretti, Magnani, and Rocco, 2004; Padthe et al., 2008).

Al-Bender and Swevers (2008) argued that the Stribeck curve applies to both liquid lubrication and dry friction. The difference is that hydrodynamic pressure plays a key role in liquid lubrication, and dry friction is controlled by the adhesion and deformation of asperities. The shape of the curve is a combination of several phenomena. 'Velocity weakening', that is, decrease of COF with velocity, is caused by the reduced time available for the deformation of asperities and formation of junctions. On the other hand, 'velocity strengthening', is attributed to 'asperity inertia effects'. A possible meaning is that asperities gain momentum with increasing speed, and are prone to larger deformation when hitting each other. It is useful to have this curve in mind when analysing the dependence of dry friction on velocity. However, we show that in quite a few cases the Stribeck curve did not apply to solid lubrication, due to the presence of phenomena absent in liquid lubrication (Section 8.1).

7.1.1 Ferrous Compounds

Given their importance in the hot rolling of steel, iron oxides are analysed in detail with other ferrous compounds that may exist in the roll gap. Regarding oxides, Bisson and co-workers (1956) conducted tests with the sliding of steel on steel without any lubrication, and with preformed oxides (Figure 7.1). The tests were carried for a wide range of velocities, and in all cases friction decreased with velocity. Magnetite was a much better lubricant than hæmatite, and the reasons are discussed in Chapter 11.

Johnson, Godfrey, and Bisson (1948) observed that COF depends on load for magnetite, but not for hæmatite (Figure 7.2), for unknown reasons. In these cases the friction generally decreases with velocity, however, it was also reported that the shape of the friction–velocity curve for magnetite can

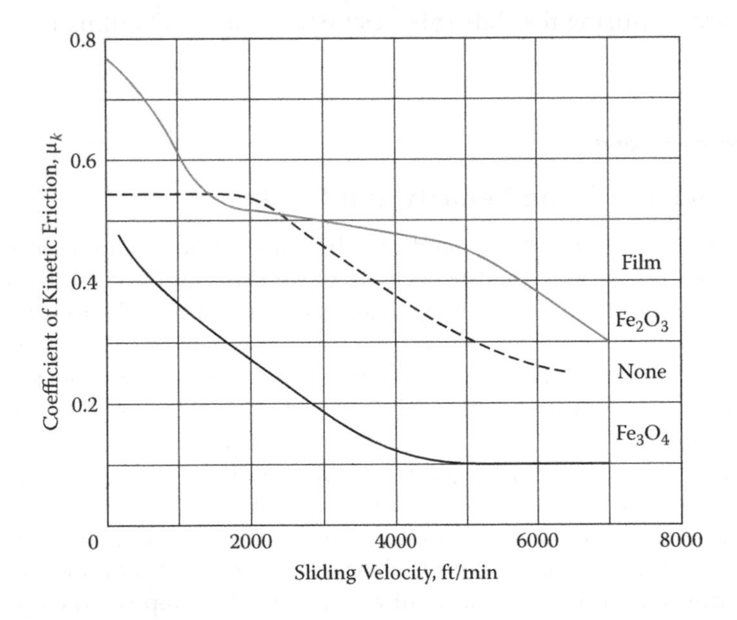

FIGURE 7.1
Friction of dry unlubricated steel against steel without film and with 1200 Å-thick films of hæmatite and magnetite. (Reprinted from E.E. Bisson, R.L. Johnson, M.A. Swikert et al. 1956, Washington: NACA Report 1254.)

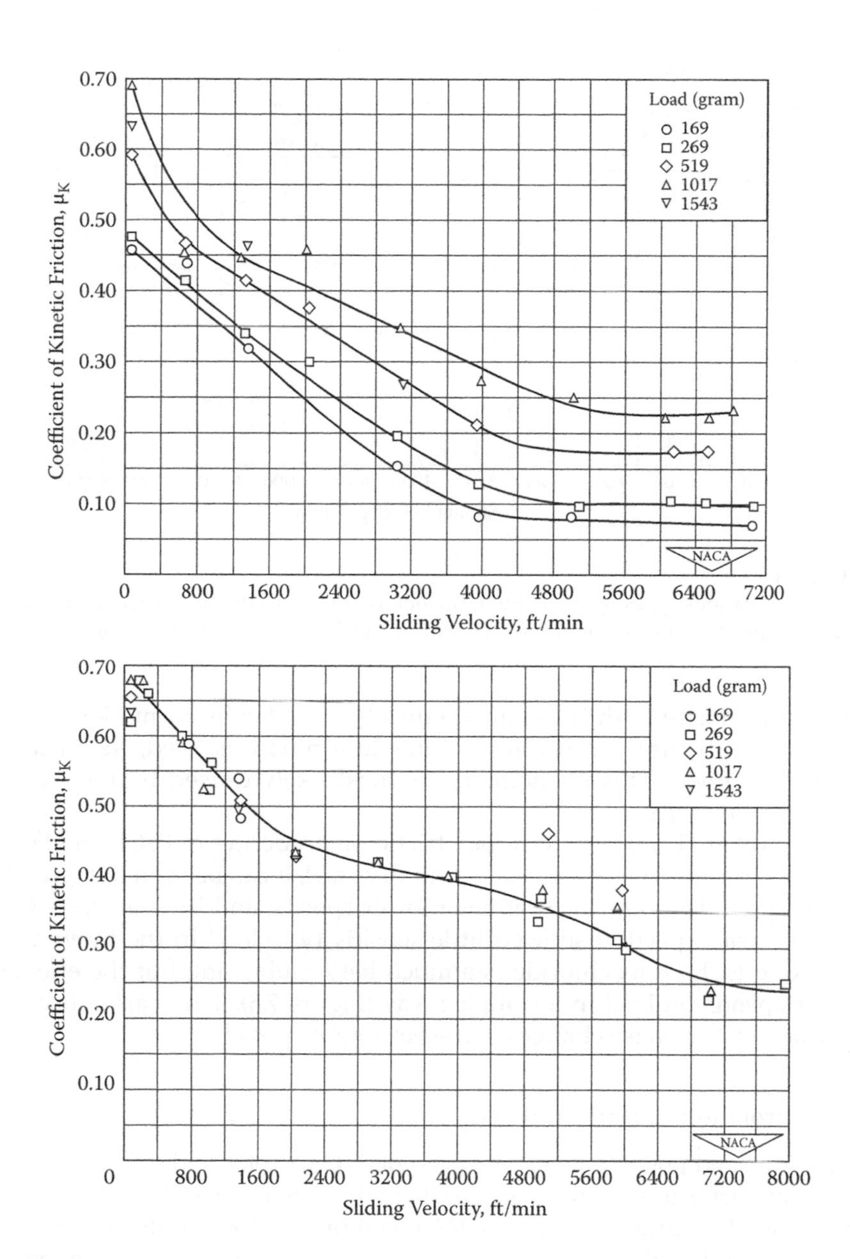

FIGURE 7.2
Effect of sliding velocity and load on friction of magnetite (top) and hæmatite (bottom). (Reprinted from R.L. Johnson, D. Godfrey, and E.E. Bisson, 1948, Washington: NACA TN 1578.)

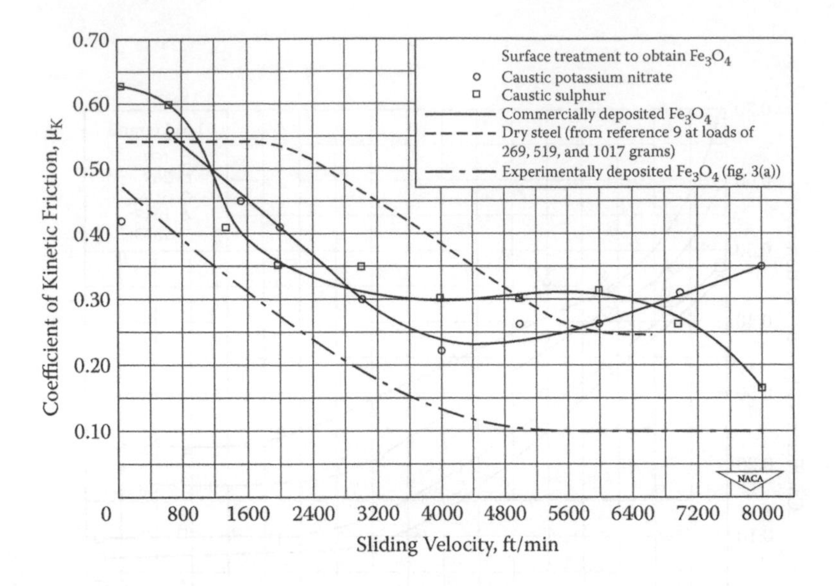

FIGURE 7.3
Effect of the various types of magnetite formation on the dependence of friction on sliding velocity. (Reprinted from R.L. Johnson, D. Godfrey, and E.E. Bisson, 1948, Washington: NACA TN 1578.)

drastically change with different chemicals used for its formation on the steel surface (Figure 7.3). If caustic potassium nitrate is used, the Stribeck curve is obtained. This phenomenon is most likely caused by the surface effects of oxide formation.

Johnson et al. (1948) also examined frictional properties of $FeCl_2$ and FeS at the steel surface. From comparison with dry steel, it can be seen in Figure 7.4 that FeS provides some lubrication at lower speeds, and in these tests COF decreases with speed. It showed little sensitivity to load in the range used in these tests. Ferrous chloride is a much better lubricant, but the effect of speed depends on load in a complex way (Figure 7.5). This variability was explained by the phase changes at the $FeCl_2$ or its possible melting.

7.1.2 Chromium Carbide and Oxide

Cr carbides, and possibly Cr_2O_3, are present in HiCr rolls, and these rolls are a significant factor in steel rolling (see Chapter 12). Sliney (1990) experimented with a solid lubricant made of 70 wt-% metal-binded Cr carbide and 15 wt-% of each Ag and CaF/BaF eutectic. The dependence is complex and depends heavily on temperature (Figure 7.6a). On the other hand, Striebing et al. (2007) experimented with PM300 (60 wt-% NiCr, 20 wt-% Cr_2O_3 Cr oxide, and 10 wt-% of each eutectic), and at 540°C observed that friction increases with speed (Figure 7.6b).

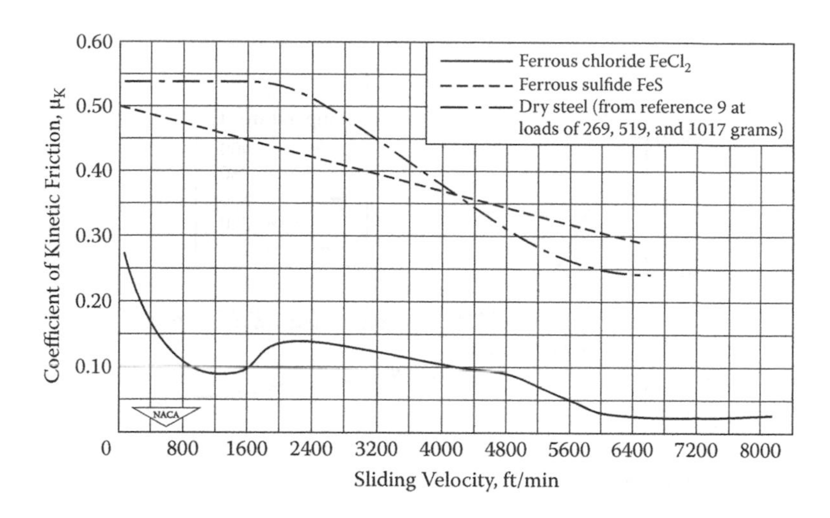

FIGURE 7.4
Effect of sliding velocity on friction of FeS and $FeCl_2$. (Reprinted from R.L. Johnson, D. Godfrey, and E.E. Bisson, 1948, Washington: NACA TN 1578.)

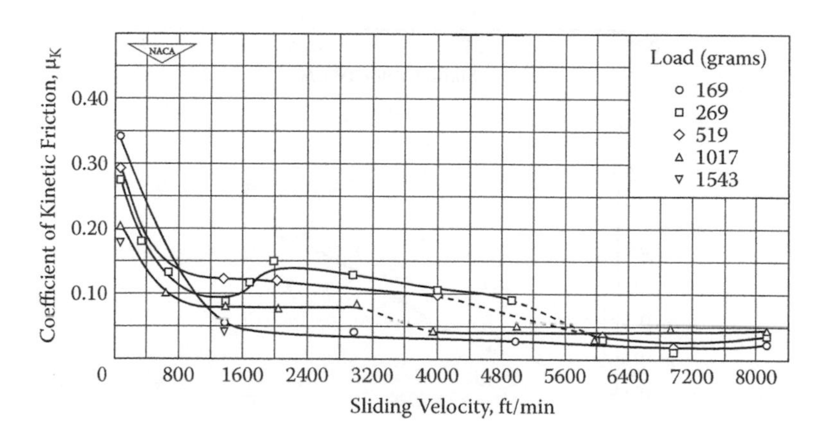

FIGURE 7.5
Effect of load and sliding velocity on friction of $FeCl_2$. (Reprinted from R.L. Johnson, D. Godfrey, and E.E. Bisson, 1948, Washington: NACA TN 1578.)

7.1.3 Metals with Phase Transformation, and Other Metallic Materials

Titanium and thallium offer insight into the friction–velocity dependence. The graphs in Figure 7.7 were obtained in vacuum, without oxidation. Buckley, Kuczkowski, and Johnson (1965) studied the sliding of Ti and its alloys over steel or itself. In most cases, friction decreased with speed until reaching a plateau. However, in one test, the titanium rider transformed its structure from hexagonal to cubic, and friction rapidly increased (Figure 7.7).

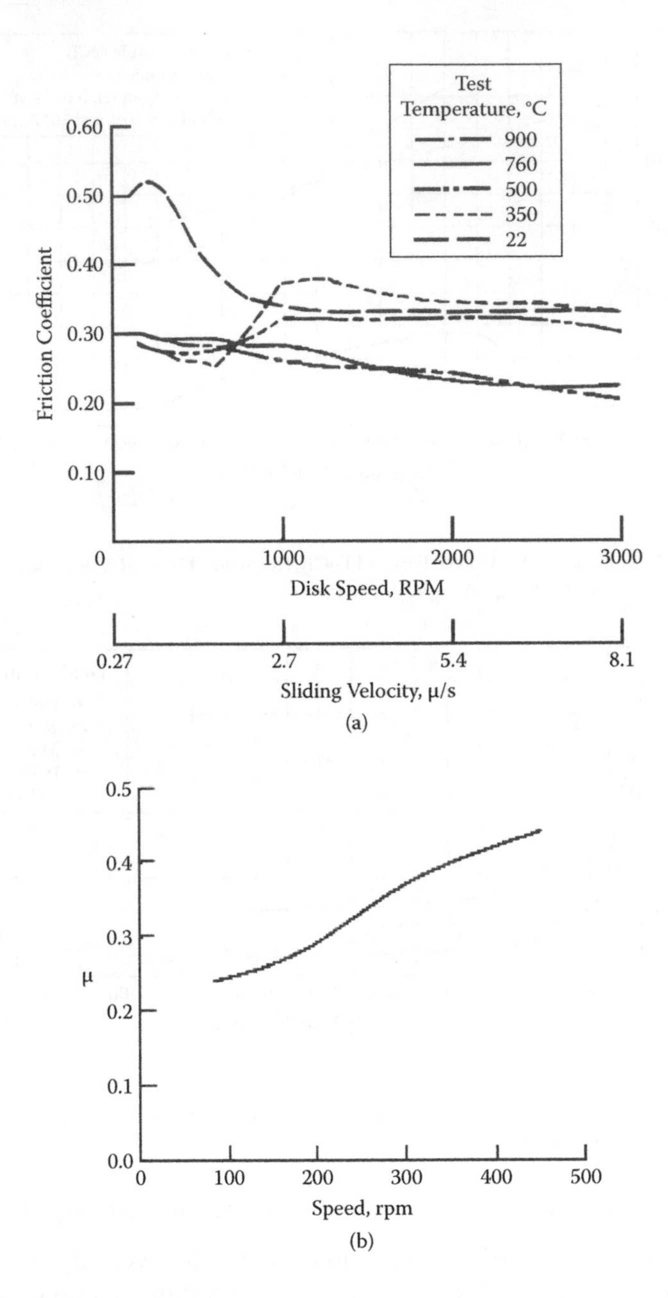

FIGURE 7.6
Effect of velocity on friction of solid lubricants PM212 and PM300 containing (a) Cr-carbide. (Reprinted from H.E. Sliney, 1990, NASA TM-103612.) (b) Cr_2O_3 (Reprinted from D.R. Striebing, M.K. Stanford, C. DellaCorte et al., 2007, NASA/TM-2007-214819.)

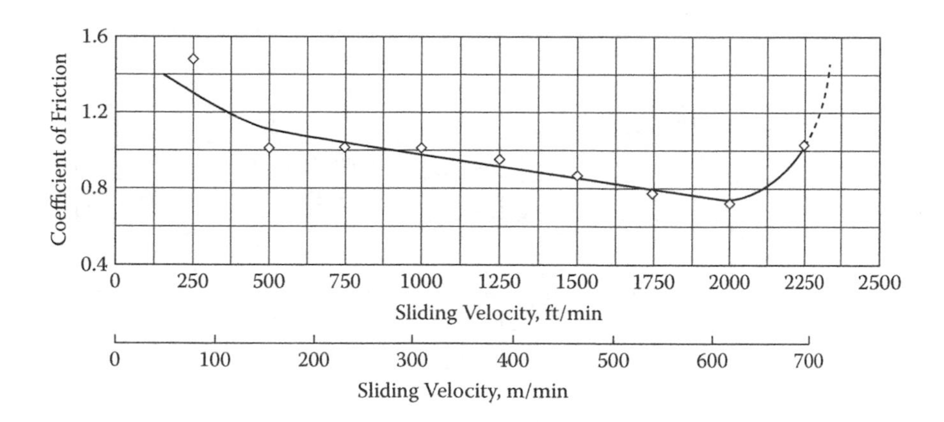

FIGURE 7.7
The impact of sliding velocity on friction of Ti-Zr alloys sliding on itself. (Reprinted from D.H. Buckley, T.J. Kuczkowski, and R.L. Johnson, 1965, NASA TN D-2671.)

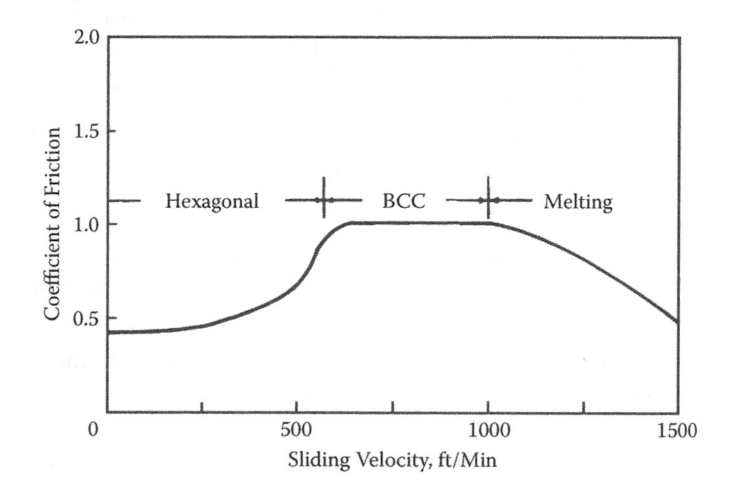

FIGURE 7.8
The impact of sliding velocity on COF of Tl on steel. (Reprinted from D.V. Keller, *13th Sagamore Conf. Physical and Chemical Characteristics of Surfaces and Interfaces*, 1966, NASA CR 82371.)

Keller (1966) noted the changes in the friction–velocity curve for Tl sliding on steel (Figure 7.8). The COF increased drastically during the transformation into a cubic structure, because metals with an hexagonal structure have lower friction than those with cubic structure (Buckley et al., 1965). After the melting point was reached, the COF decreased with speed. In both cases, Ti and Tl, the transformation was brought about by the temperature increase due to frictional heating.

Regarding other metallic materials, the experiments by Johnson, Swikert, and Bisson (1952), summarised in Figure 7.9, show a bewildering variety of

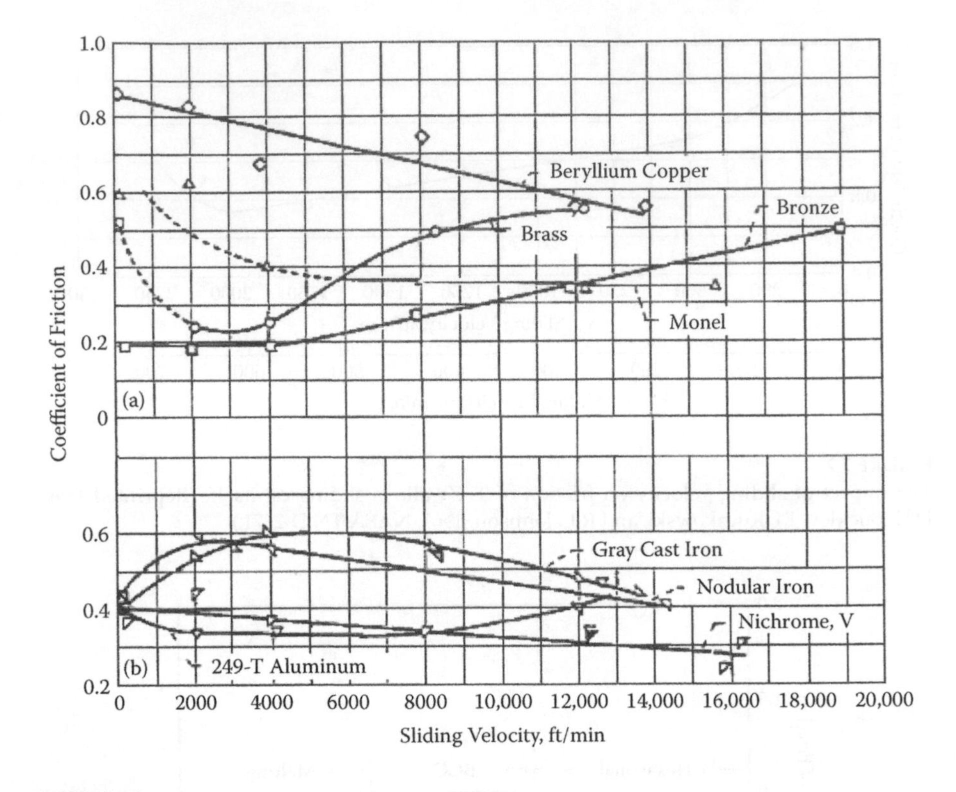

FIGURE 7.9
Effect of sliding velocity on friction of several metallic materials sliding on steel without lubrication. (Reprinted from R.L. Johnson, M.A. Swikert, and E.E. Bisson, 1952, Washington: NACA TR-1062.)

the shapes of friction–velocity dependence. The authors contended that friction increases with velocity after oxide on the surface fails.

7.2 Various Explanations of Friction–Velocity Dependence

Various theories were proposed to explain the dependence of friction on velocity:

1. According to Johnson, Swikert, and Bisson (1947), COF decreases with velocity because increased frictional heating warms up the surface, reducing its shear strength. Also, at the higher surface temperature of steel, FeO formed, which coincided with the onset of COF reduction.

2. Similarly, Bisson et al. (1956) asserted that the decrease in friction with velocity is caused by oxidation, enhanced at high velocities by frictional heat.
3. Forrester (1946) argued that an increase in friction with velocity is due to a partial destruction of the boundary film, the rate of destruction growing with increasing sliding speed; a similar theory was also proposed by Johnson et al. (1952). However, they also argued that increasing velocity increases surface temperature, which reduces the strength of the junctions of asperities. This effect is much more pronounced in materials of low melting point. However, wear (which may be accelerated at higher velocity) increases the contact area, reducing the contact pressure and local temperature.

The first two theories are supported by observations of the effect of speed on the proportion of iron oxides found in wear debris generated by steel sliding (Quinn, 1991). Generally, a portion of Fe_2O_3 decreases with speed, and the portions of FeO and F_3O_4 increase. As shown in Section 13.3, magnetite and wüstite are much softer tha hæmatite, and can act as lubricants on the steel surface.

7.3 Some Specific Aspects of Solid Lubrication

From the simplified modelling in Section 4.4 it follows that placing a low-shear-strength film between hard materials reduces friction. The film separates the hard surfaces and prevents their direct contact and adhesion. Less force is required to shear junctions between the film and the base, or two films (as is the case in hot rolling, with oxide on both the strip and the roll), than between metals. Bowden and Tabor (1950) pointed out that shearing takes place within the bulk of softer material. The impact of shear strength is illustrated in Figure 7.10. It should be noted here that the terms hardness and shear strength are often interchangeable in the literature. In principle, there is a positive, although not always linear, correlation between hardness, and the shear and tensile strengths.

However, if the film is much harder than the base, the base will break under heavy load, the film may rupture and adhesion can occur, as at the Al/Al_2O_3 interface (Bowden and Tabor, 1950). If a metal and film have similar hardness, the film deforms with the underlying metal and does not break. Oxide, though, can be somewhat harder and still act as a lubricant, as shown for copper.

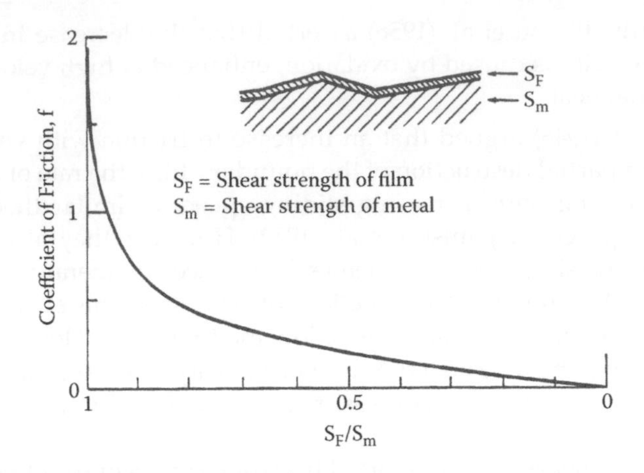

FIGURE 7.10
COF as a function of the ratio of shear strengths of film and metal. (Reprinted from D. Godfrey, 1968, Boundary lubrication. In *Interdisciplinary Approach to Friction and Wear,* P.M. Ku (Ed.), 335–384, NASA SP-181.)

References

Al-Bender, F., and J. Swevers. 2008. Characterization of friction force dynamics. *IEEE Control Sys. Mag.* 28(6):64–81.

Åström, K.J., and C. Canudas-de-Wit. 2008. Revisiting the LuGre friction model. *IEEE Control Sys. Mag.* 28(6):101–114.

Bisson, E.E., R.L. Johnson, M.A. Swikert et al. 1956. *Friction, Wear and Surface Damage of Metals as Affected by Solid Surface Films.* Washington: NACA Report 1254.

Bowden, F.P., and D. Tabor. 1950. *The Friction and Lubrication of Solids.* Oxford: Oxford University Press.

Buckley, D.H., T.J. Kuczkowski, and R.L. Johnson. 1965. *Influence of Crystal Structure on Friction and Wear of Titanium and Titanium Alloys in Vacuum.* NASA TN D-2671.

Canudas-de-Wit, C., H. Olsson, K.J. Åström et al. 1995. A new model for control of systems with friction. *IEEE Trans. AC* 40:419–425.

DellaCorte, C. 1987. *Experimental Evaluation of Chromium-Carbide-Based Solid Lubricant Coating for Use to 760°C.* NASA CR 180808.

Ferretti, G., G. Magnani, and P. Rocco. 2004. Single and multistate integral friction models. *IEEE Trans. AC* 49:2292–2297.

Forrester, P.G. 1946. Kinetic friction in or near the boundary region. II. The influence of sliding velocity and other variables on kinetic friction in or near the boundary region. *Proc. R. Soc. Lond.* A 187:439–463.

Godfrey, D. 1968. Boundary lubrication. In *Interdisciplinary Approach to Friction and Wear,* P.M. Ku (Ed.), 335–384. NASA SP-181.

Hironaka, S. 1984. Boundary lubrication and lubricants. *Three Bond Technical News.*

Johnson, R.L., M.A. Swikert, and E.E. Bisson. 1947. *Friction at High Sliding Velocities.* Washington: NACA TN 1442.

Johnson, R.L., D. Godfrey, and E.E. Bisson. 1948. *Friction of Solid Films on Steel at High Sliding Velocities*. Washington: NACA TN 1578.

Johnson, R.L., M.A. Swikert, and E.E. Bisson. 1952. *Investigation of Wear and Friction Principles Under Sliding Conditions of Some Materials Suitable for Cages of Rolling-Contact Bearings*. Washington: NACA TR-1062.

Keller, D.V. 1966. Application of recent static adhesion data to the adhesion theory of friction. In *13th Sagamore Conference on Physical and Chemical Characteristics of Surfaces and Interfaces*. NASA CR 82371.

Kelly, R., J. Llamas, and R. Campa. 2000. A measurement procedure for viscous and Coulomb friction. *IEEE Trans. Instr. Meas*. 49:857–861.

Padthe, A.K., B. Drincic, J. Oh et al. 2008. Duhem modelling of friction-induced hysteresis. *IEEE Con. Sys. Mag*. 28(5):90–107.

Quinn, T.F.J. 1991. *Physical Analysis for Tribology*. Cambridge: Cambridge University Press.

Sliney, H.E. 1987. Self-lubricating coatings for high-temperature applications. In *Aeropropulsion '87*, 89–101.

Sliney, H.E. 1990. *Composite Bearing and Seal Materials for Advanced Heat Engine Applications to 900°C*. NASA TM-103612.

Striebing, D.R., M.K. Stanford, C. DellaCorte et al. 2007. *Tribological Performance of PM300 Solid Lubricant Bushings for High Temperature Applications*. NASA/TM-2007-214819.

Thomsen, J.J. 1999. Using fast vibrations to quench friction-induced oscillations. *J. Sound Vib*. 228:1079–1102.

8

Modelling of Friction in Control Engineering

These models are considered before the general models because they are more articulate and introduce important concepts clearly. Quite a few control engineers in academia specialise in friction, and there are concrete applications of friction models to the positioning of telescopes (Nurre, 1974; Rivetta and Hansen, 1998), robots (Swevers et al., 2000), precise machining (Dupont et al., 2002), vibrations in oil drilling (Kyllingstad and Halsey, 1988), tyre-road contact (Canudas-de-Wit et al., 2003), wafer polishing in semiconductor manufacturing (Yi, 2008) and railway braking (Park et al., 2008). Several good reviews of the friction modelling in control engineering have been published (Armstrong-Hélouvry, Dupont, and Canudas-de-Wit 1994; Andersson, Söderberg, and Björklund, 2007; Armstrong and Chen, 2008; Wojewoda et al., 2008).

In terms of the classification of models, this section borrows from the succinct review by Iurian et al. (2005). They classify the models into static and dynamic. 'Static' does not mean static friction, but the absence of frictional memory, where only the present conditions determine future trends. Dynamic models use both past and present frictional behaviour to predict the future. The classification by Al-Bender, Lampaert, and Swevers (2005) is somewhat different

1. *Physically motivated models.* The behaviour of a sliding asperity is modelled using first principles. The focus is on the micro- and nanoscale, and the researchers generally do not attempt to extrapolate the models to macroscopic scale.
2. *Empirical models.* These consider the macroscopic scale, and are obtained by fitting tuning factors to observations.

8.1 Static Friction Models

8.1.1 Basic Models and Their Shortcomings

There are four basic types of these steady-state models, where COF is a function of velocity:

1. The simplest form is so-called Coulomb friction, with COF of constant magnitude (although Coulomb did not advocate independence of friction from velocity).
2. Viscous friction is expressed as a linear function of sliding velocity, v_S, $\mu = \mu_0 + a\, v_S$.
3. Viscous friction can be extended with static friction (*stiction*), where at zero velocity COF is greater than parameter μ_0 in the equation above.
4. The previous case can be modified by the inclusion of the Stribeck curve for $v_S > 0$.

Some phenomena cannot be accurately addressed by these models (Canudas-de-Wit et al., 1995; Iurian et al., 2005):

1. *Pre-sliding.* Courtney-Pratt and Eisner (1957) observed pre-sliding (i.e., a displacement that occurs before sliding commences) at the tangential force smaller than the static friction force. Their interpretation was that adhering asperities are first deformed elastically, then plastically, and if the stiction force is reached, the asperity junctions break and sliding starts. The plastic deformation also leads to another frictional phenomenon, the hysteresis loop (Figure 8.1a).
2. *Friction lag.* Due to the hysteresis (Figure 8.1b), friction is different for increasing and decreasing velocity, at the same velocity magnitude (Canudas-de-Wit et al., 1995).
3. *Varying breakaway force.* The breakaway force is the force required to overcome static friction and initiate sliding, and it decreases with the increasing rate of load change.
4. *Stick-slip motion.* Canudas-de-Wit and co-workers (1995) illustrated this with a system on a flat surface consisting of a block connected to a spring, the other end of which is pulled at constant speed. The block is initially at rest, and will start to move when the spring force reaches the breakaway force. The mass slides and friction decreases, the block accelerates, and the spring contracts. The spring force decreases, the block slows down, friction increases, and the block eventually stops. The cycle then repeats.

8.1.2 Seven-Parameter Model

This model addressed all the phenomena above (Armstrong-Hélouvry et al., 1994), although Iurian et al. (2005) still classified it as static, inasmuch as the frictional memory, that is, the frictional behaviour in the past, was implemented as a time lag. There are other shortcomings. First, Canudas-de-Wit et al. (1995) commented that it does not combine various phenomena, but consists of separate models for stiction and sliding. Second, switching of the

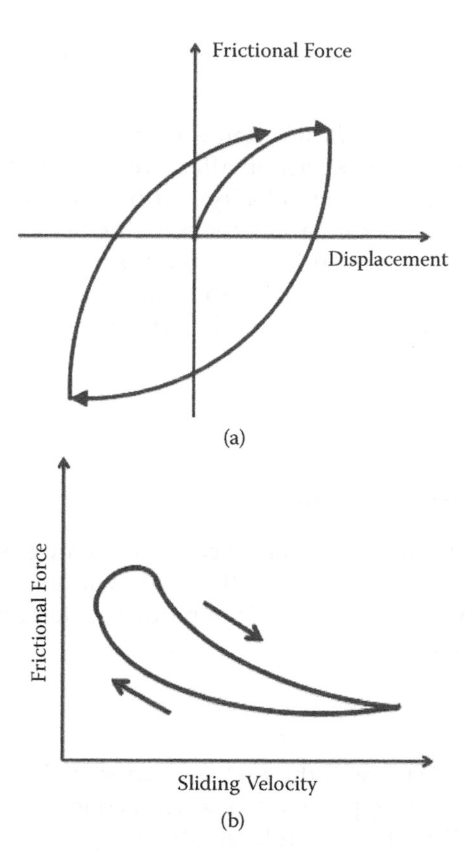

FIGURE 8.1
Some phenomena that cannot be accurately modelled with static models: (a) hysteresis loop, and (b) friction lag.

model between the kinetic and the static friction is complicated, because it requires the initialisation of position and velocity every time the switch occurs (Iurian et al., 2005).

The instantaneous friction force F_f for the pre-sliding displacement x is given by Equation (8.1), and for the sliding (which combines Coulomb and viscous friction, and the Stribeck curve with frictional memory), by Equation (8.2). The notion that static friction may increase with the dwell time at rest is also included [Equation (8.3)].

$$F_f(x) = -k_t x \tag{8.1}$$

$$F_f(v,t) = -\left\{ \frac{F_C + F_v|v| + F_S(\gamma, t_2)}{1 + \left(\frac{v(t-\tau)}{v_S} \right)^2} \right\} \operatorname{sgn}(v) \tag{8.2}$$

$$F_S(\gamma, t_2) = F_{S,a} + (F_{S,\infty} - F_{S,a})t_2 / (t_2 + \gamma) \quad (8.3)$$

The seven parameters are F_C (Coulomb friction), F_v (viscous friction), $F_{s,\infty}$ (the magnitude of the Stribeck friction after a rest), k_t (the tangential stiffness of static contact), v_s (the sliding velocity), τ (the time constant of frictional memory) and γ (the temporal parameter of rising static friction). The other symbols above are F_s, the magnitude of the Stribeck friction (NB: the frictional force at breakaway is $F_C + F_s$), $F_{s,a}$, the magnitude of the Stribeck friction at the end of the previous sliding period, t_2, the dwell time, and t, time.

8.2 Dynamic Friction Models

Dynamic models eliminate the complicated switching between static and kinetic friction. They introduce so-called state variables, which are linked to the magnitude of friction, and their dynamics is described by differential equations (Iurian et al., 2005).

8.2.1 Dahl

The model by Dahl (1968) was the first friction model with a state variable. Its basis is the adhesion hypothesis; asperities contact each other and form bonds. During sliding, bonds undergo shearing stresses. If the displacement is large, yielding starts first, followed by rupture. It is claimed that static friction and Coulomb friction are indistinguishable for materials that exhibit brittle fracture; that is, once the maximum stress is reached, the bond is broken. On the other hand, ductile materials pass through the 'tacky' stage between the maximum stress and the rupture. Its mathematical formulation was not articulated in the original paper, and a more comprehensive representation was offered by Dupont et al. (2002), where the displacement x is a sum of the elastic and plastic components z and w, respectively, although the latter is represented implicitly [Equation (8.4)]. If σ_0 is the 'contact stiffness', the friction stress is represented as:

$$f = \sigma_0 z, \quad \sigma_0 > 0; \quad \dot{z} = \dot{x}\left(1 - \frac{\sigma_0}{F_C}\operatorname{sgn}(\dot{x})z\right) \quad (8.4)$$

8.2.2 LuGre

Researchers from the universities in Lund and Grenoble (hence LuGre) combined the Dahl model with some steady-state friction features, including the

Stribeck curve (Canudas-de-Wit et al., 1995; Åström and Canudas-de-Wit, 2008). The basic idea was that rigid bodies contact through elastic bristles. Tangential force deflects bristles like springs, producing frictional force. When the force exceeds a threshold, the deflection of bristles is so large that they slip. A mathematical formulation was proposed by Dupont et al. (2002) as:

$$f = \sigma_0 z + \sigma_1 \dot{z} + \sigma_1 \dot{x}, \quad \sigma_0, \sigma_1, \sigma_2, > 0 \tag{8.5}$$

$$\dot{z} = \dot{x}\left(1 - \frac{\sigma_0}{\left|f_{ss}(\dot{x})\right|}\operatorname{sgn}(\dot{x})z\right) \tag{8.6}$$

where z is the average deflection of bristles, σ_0 the bristle stiffness, σ_1 the bristle damping coefficient, σ_2 the viscous damping coefficient, and f_{SS} represents the Stribeck curve. The LuGre model can be reduced to Dahl's model if $\sigma_1 = \sigma_2 = 0$, and $f_{ss} = F_C$.

Many applications have been reported, and LuGre is credited as the first model with a smooth switch between stiction and sliding (Tjahjowidodo et al., 2007). Swevers et al. (2000) praised its elegance and easy implementation, and it performed well in experimental validation (Section 8.3). There are some criticisms, however. Park et al. (2008) contended that it has too many tuning parameters; Dupont et al. (2002) noted that it does not handle friction force with a small vibrational component. Swevers et al. (2000) claimed (a) that some aspects of hysteretic behaviour are not accounted for, and (b) the model cannot be easily adapted to measured values. They proposed the so-called Leuven model, described below.

8.2.3 Leuven

Swevers and co-workers proposed the Leuven model, actually named by Lampaert, Swevers, and Al-Bender (2002):

$$\dot{z} = v\left[1 - \operatorname{sgn}\left(\frac{F_d(z)}{s(v) - F_o}\right)\left|\frac{F_d(z)}{s(v) - F_o}\right|^n\right] \tag{8.7}$$

$$F_d = F_h(z) + \sigma_1 z + \sigma_2 v \tag{8.8}$$

where v is the current velocity, n is the coefficient to tune the shape of the force-position relationship, and $s(v)$ is a function that models the constant velocity behaviour:

$$s(v) = \operatorname{sgn}(v)\left[F_C + (F_0 - F_C)e^{-(|v|/V_s)^{\delta}}\right] \tag{8.9}$$

where F_o is the static friction and δ an arbitrary exponent. $F_h(z)$ is the hysteresis force, σ_1 is the microviscous damping coefficient and σ_2 the viscous damping coefficient.

8.2.4 Generalised Maxwell Slip

The authors of the Leuven model noted two shortcomings, namely the discontinuity in the friction force upon closing a hysteresis loop, and numerical problems in the implementation of the hysteresis model (Tjahjowidodo et al., 2007). They devised the generalised Maxwell slip model (GMS; Al-Bender, Lampaert, and Swevers 2004; Al-Bender, Lampaert, and Swevers 2005). It uses a parallel system of blocks and springs (Figure 8.2), where each block-and-spring element is described with a state equation (Tjahjowidodo et al., 2007). The friction force is estimated as a sum of the outputs from all blocks. Each block is a generalised asperity, which can either stick or slip, and is treated as in Figure 8.3. It is assumed that there are three basic mechanisms of friction, namely (a) the creep of the contacting asperities in the normal direction, (b) adhesion, and (c) deformation (Al-Bender et al., 2004).

8.2.5 Dupont

Dupont et al. (2002) contended that the modelling of pre-sliding as a combination of elastic and plastic displacement exhibits a non-physical phenomenon

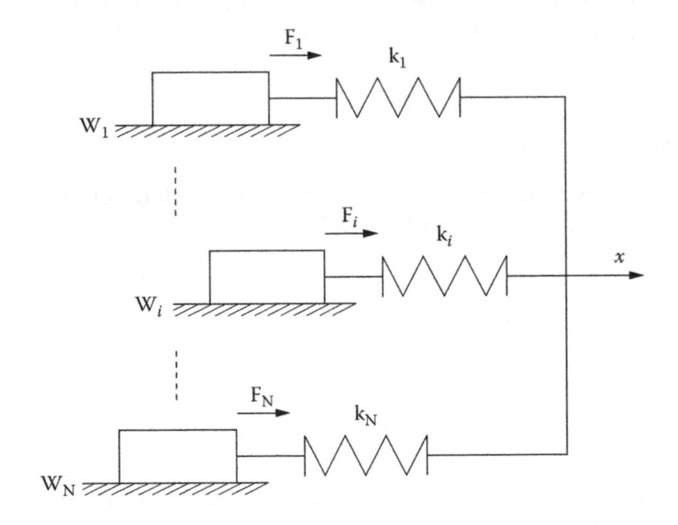

FIGURE 8.2
Representation of GMS with N elementary models. (Reprinted from T. Tjahjowidodo et al., *J. Sound Vib.* 308:632–646, 2007. With permission from Elsevier.)

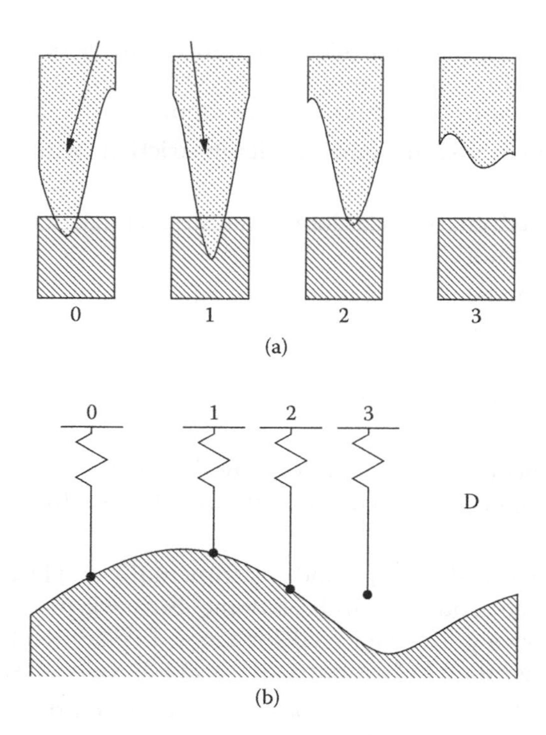

FIGURE 8.3
Representation of (a) individual asperities enumerated 0–3, (b) in contact with a substrate, using GMS. (Reprinted from F. Al-Bender, V. Lampaert, and J. Swevers, *Tribol. Lett.* 16:81–93, 2004. With permission from Springer Verlag.)

called drift, where a block placed in frictional contact with an inclined plane and subjected to small vibrations creeps down. This is not observed, and the proposed 'elastoplastic pre-sliding' assumes that the bond between asperities is initially elastic. Further stretching makes the bond deformation first a mixture of plastic and elastic, then fully plastic, until the bond ruptures and sliding begins. It is reported that the proposed model substantially reduces the drift.

8.3 Experimental Validation of Models

The friction models were validated experimentally in several cases, and dynamic models outperformed the static ones. The good performance of LuGre and GMS is noteworthy:

1. Kelly, Llamas, and Campa (2000) measured the velocity response of a drive rotor to a torque ramp. The experimental observations were then compared to the results of three models:
   a. Combined viscous and Coulomb friction without frictional memory.
   b. Combined viscous, Coulomb and Stribeck friction without frictional memory.
   c. LuGre model. A good qualitative agreement was obtained only for this model.
2. Padthe et al. (2008) tested three friction models (Dahl, LuGre and GMS) on an experimental test bed regarding hysteresis, and LuGre performed best.
3. Yi (2008) incorporated the LuGre model in the model of the polishing of semiconductor wafers, the results of which agreed well with measurements.
4. Tjahjowidodo et al. (2007) found experimentally that LuGre and GMS outperformed Coulomb and Coulomb–Stribeck models, although the latter ones performed well for large displacements. GMS performed better than LuGre, particularly at low velocities.
5. Park et al. (2008) obtained a good match between the experimental observations and modelling results for their model combining the Dahl model with the Stribeck curve.

References

Al-Bender, F., V. Lampaert, and J. Swevers. 2004. A novel generic model at asperity level for dry friction force dynamics. *Tribol. Lett.* 16:81–93.

Al-Bender, F., V. Lampaert, and J. Swevers. 2005. The generalised Maxwell-slip model: A novel model for friction simulation and compensation. *IEEE Trans. AC* 50:1883–1887.

Andersson, S., A. Söderberg, and S. Björklund. 2007. Friction models for sliding dry, boundary and mixed lubricated contacts. *Tribol. Int.* 40:580–587.

Armstrong-Hélouvry, B., P. Dupont, and C. Canudas-de-Wit. 1994. A survey of models, analysis tools and compensation methods for the control of machines with friction. *Automatica* 30:1083–1138.

Armstrong, B.S.R., and Q. Chen. 2008. The z-properties chart. *IEEE Control Sys. Mag.* 28(5):79–89.

Åström, K.J. and C. Canudas-de-Wit. 2008. Revisiting the LuGre friction model. *IEEE Con. Sys. Mag.* 28 (6): 101-114.

Canudas-de-Wit, C., H. Olsson, K.J. Åström et al. 1995. A new model for control of systems with friction. *IEEE Trans. AC* 40:419–425.

Canudas-de-Wit, C., P. Tsiotras, E. Velenis et al. 2003. Dynamic friction models for road/tire longitudinal interaction. *Vehicle Sys. Dyn.* 39:189–226.

Courtney-Pratt, J.S., and E. Eisner. 1957. The effect of a tangential force on the contact of metallic bodies. *Proc. R. Soc. Lond. A* 238:529–552.

Dahl, P.R. 1968. *A Solid Friction Model.* El Segundo, CA: Aerospace Corp., Rep. TOR-0158 (3107-18)-1.

Dupont, P., V. Hayward, B. Armstrong, et al. 2002. Single-state elastoplastic friction models. *IEEE Trans. AC* 47:787–792.

Iurian, C., F. Ikhouane, J. Rodellar et al. 2005. *Identification of a System with Dry Friction.* Universitat Politècnica de Catalunya, Institut d'Organització i Control de Sistemes Industrials: IOC-DT-P-2005-20.

Kelly, R., J. Llamas, and R. Campa. 2000. A measurement procedure for viscous and Coulomb friction. *IEEE Trans. Instr. Meas.* 49:857–861.

Kyllingstad, Å., and G.W. Halsey. 1988. A study of slip/stick motion of the bit. *SPE Drilling Eng.* 3:369–373.

Lampaert, V., J. Swevers, and F. Al-Bender. 2002. Modification of the Leuven integrated friction model structure. *IEEE Trans. AC* 47:683–687.

Nurre, G.S. 1974. *An Analysis of the Dahl Friction Model and Its Effect on a CMG Gimbal Rate Controller.* NASA TM X-64934.

Padthe, A.K., B. Drincic, J. Oh et al. 2008. Duhem modelling of friction-induced hysteresis. *IEEE Con. Sys. Mag.* 28(5):90–107.

Park, S.H., J.S. Kim, J.J. Choi et al. 2008. Modelling and control of adhesion force in railway rolling stocks. *IEEE Control Sys. Mag.* 28(5):44–58.

Rivetta, C.H., and S. Hansen. 1998. *Friction Model of the 2.5mts SDSS Telescope.* Fermi National Accelerator Laboratory: FERMILAB-Conf-98/070.

Swevers, J., F. Al-Bender, C.G. Ganseman et al. 2000. An integrated friction model structure with improved presliding behaviour for accurate friction compensation. *IEEE Trans. AC* 45:675–686.

Tjahjowidodo, T., F. Al-Bender, F., H. Van Brussel et al. 2007. Friction characterisation and compensation in electro-mechanical systems. *J. Sound Vib.* 308:632–646.

Wojewoda, J., A. Stefanski, M. Wiercigroch et al. 2008. Hysteretic effects of dry friction: Modelling and experimental studies. *Phil. Trans. R. Soc. A* 366:747–765.

Yi, J. 2008. Friction modelling in linear chemical-mechanical planarization. *IEEE Control Sys. Mag.* 28(5):59–78.

9

Modelling of Macroscopic Friction

Friction models based on first principles are far from mature. Unlike, say, heat transfer and fluid dynamics, tribology is not very amenable to mathematical modelling. Ludema (1996) noted that over the then preceding 40 years more than 300 equations for the prediction of friction had been published, 'but even the best has very limited use', the key reasons being:

1. Most authors select a very narrow range of variables and then assume that friction varies monotonically over these ranges.
2. Friction mechanisms are assumed invariant in time, that is, not affected by past events.
3. Friction is measured over a limited range of selected variables, and the data are then fitted into an equation, without much regard for the underlying mechanisms.
4. The chosen variables in an equation are treated as mutually independent.

He illustrated these issues with the wear rate of a 60 Cu–40 Zn pin on a high-speed steel ring measured over a wide range of temperature and sliding speed (Figure 9.1). A simple curve fitting within different narrow ranges of speed or temperature would produce very different models. He recommended that improved modelling requires:

1. Measurement of friction and wear over a wide range of values of all relevant variables
2. Description of testing conditions and the properties of equipment and materials, for consistent comparison with the literature, and assessment of the limits of the models based on those data
3. Explicit modelling of all relevant friction and wear mechanisms in the given case
4. Multidisciplinary approach

Ferrante (1988) asserted that: 'At this point there isn't a sufficient data base obtained under well controlled conditions to develop theoretical models for fundamental tribology'. It is interesting that Kragelsky (from Section 4.3.3) published a sizeable monograph on the calculation of friction and wear

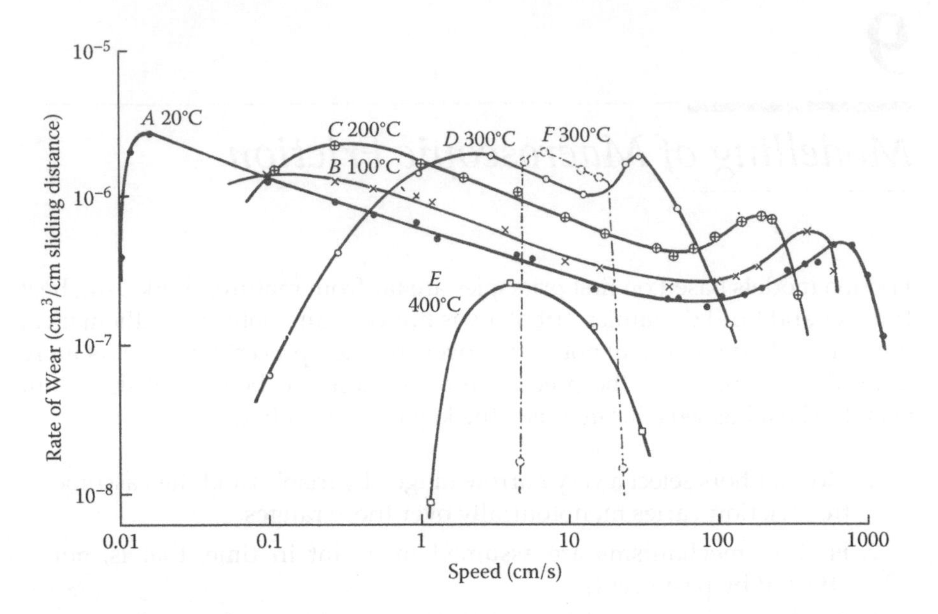

FIGURE 9.1
Wear versus sliding speed and temperature. (Reprinted from J.K. Lancaster, *Proc. R. Soc. A* 273:466–483, 1963. With permission from the Royal Society.)

(Kragelsky, I.V., M.N. Dobychin, and V.S. Kombalov. 1977. *Fundamentals of the Calculations of Friction and Wear.* Moscow: Mashinostroenie). However, the friction-related part is mainly about the calculations of details such as the size of surface contacts, asperity deformations and so on, and is of little use to practitioners. Despite all this, some genuine progress has been made. First we present the work of Ernst and Merchant and some recent variations to it, followed by the modelling by Straffelini.

9.1 Ernst and Merchant, with Recent Variations

In 1940, Ernst and Merchant provided models of friction in metal cutting (Kragelsky and Shchedrov, 1956; Bisson, 1972), where the coefficient of static friction depends on both adhesion and roughness:

$$\mu = S/H_B + \tan\varphi \tag{9.1}$$

S is the averaged shear strength over the real contact area and H_B hardness by Brinell [*cf.* Equation (4.7)]. For metals, φ (the angle between the asperities and the frictional force) is assumed negligible. Shear strength of a crystalline solid was calculated as a third of the pressure required to melt the substance at the temperature T [K]; that is,

$$S = 0.427(L/3)\rho \ln(T_m/T) \tag{9.2}$$

where T_m is the melting temperature of the solid (K), ρ its density (g cm^{-3}) and L the latent heat of fusion of the solid (cal g^{-1}). If different metals are sliding over each other, the shear strength of the metal with the lower melting temperature and the hardness of softer metal are used. Ernst and Merchant compared the calculated and measured values of COF for two groups of metal combinations, namely pairs forming solid solutions at room temperatures, and pairs almost mutually insoluble at high temperature. The agreement was quite good. Merchant concluded that the pairs of metals that do not form a solid solution have low friction, due to the lower mutual adhesion, which agrees with the observations by other researchers (Section 10.2).

Halling (1982) devised a more general model, addressing a film-covered surface sliding on a rough surface. The film can be solid or liquid, and the ploughing was neglected. There are three basic situations shown in Figure 9.2:

a. With a thin layer, body (3) is in contact both with the film (2) and the body (1) carries the full load, W.
b. With a thicker layer, a part of the load will be carried out by the film, and the rest by the body (1).
c. With a very thick layer, the load will be carried out entirely by the film.

Halling assumed that both the layer and the substrate deform plastically, and proposed the following generic formula for the coefficient of friction:

$$\mu = (\tau_1 A_1 + \tau_2 A_2)/(p_1 A_1 + p_2 A_2) \tag{9.3}$$

where subscripts 1 and 2 denote contacts between surfaces 1 and 2, and 2 and 3, respectively, p is pressure, A surface area and τ shear stress. Another interesting feature is the analysis of the effects of film thickness in the common case of plastic films on plastic surfaces, which was supported by experimental results. With soft film on a harder substrate, friction decreases with film thickness, passes through a minimum and rises again (Figure 9.3a). The opposite trend was observed with hard layer on a soft substrate (Figure 9.3b).

These results were not explained using first principles. Although Figure 9.3a can be explained using the argument in Section 4.4.1.3, the pattern in Figure 9.3b cannot be easily interpreted. Pasaribu (2005) added the impact of ploughing to the model by Halling, and compared the predictions of the two models to measurements. Both models demonstrated respectable performance, with Pasaribu's model performing slightly better.

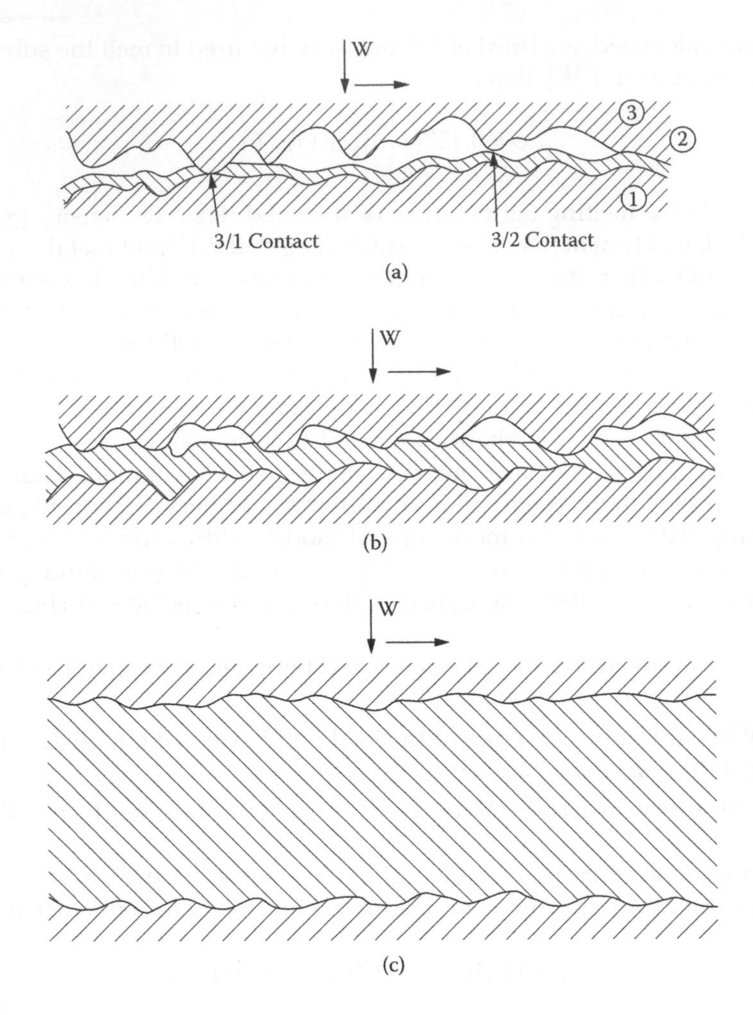

FIGURE 9.2
General model for the surface contact in the presence of film. (Reprinted from J. Halling, *ASLE Trans.* 25:528–536, 1982. With permission from Taylor & Francis.)

9.2 Straffelini, and the Work of Adhesion

Following the theory that friction is due to the formation of junctions by adhesion and the subsequent breaking of the junctions, Straffelini (2001) based his approach on the thermodynamic work of adhesion:

$$W_{ab} = \gamma_a + \gamma_b - \gamma_{ab} \tag{9.4}$$

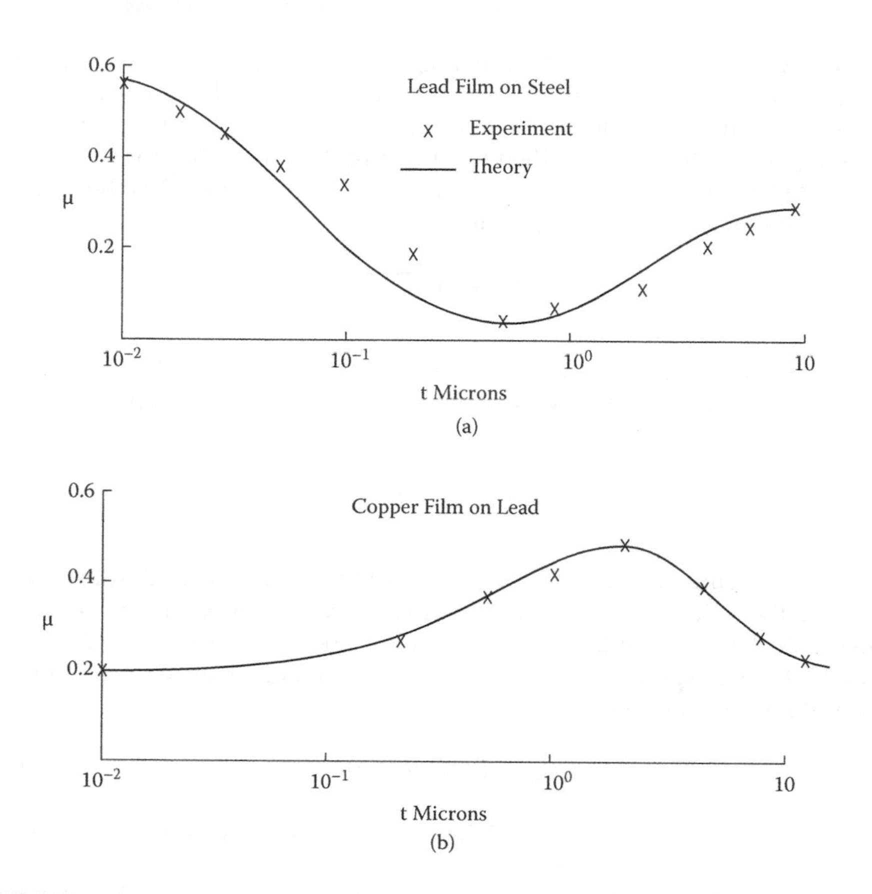

FIGURE 9.3
Effect of solid film thickness on friction for (a) soft film on hard substrate, and (b) hard film on soft substrate. (Reprinted from J. Halling, *ASLE Trans*. 25:528–536, 1982. With permission from Taylor & Francis.)

where γ_a and γ_b are the surface energies of two surfaces, and γ_{ab} is the interface energy. For metals, the work is strongly influenced by contaminants, and they tend to reduce it, although their impact is not modelled in the paper. In the derivation of the formula for the coefficient of friction, the surface energies and the interface energy are not considered. Instead, correlation between the coefficient of friction and the work of adhesion was established as:

$$\mu/\sqrt{1+12\mu^2} = 0.127W_{ab} \tag{9.5}$$

As seen in Section 4.8, the agreement between the calculated and measured COF was very good for a number of metal pairs. Jupp, Talamantes-Silva, and Beynon (2004) modified Straffelini's model to suit the hot rolling process with a thin layer of adherent magnetite in the roll gap. This required the addition of temperature effects and a method of calculating adhesion work, including oxides present in hot rolling. Using additional experimental data, Equation (9.6) was modified by changing the multiplier to 0.164. They contended that the interface energy is not as accurately defined as the surface energy, and used the following formula:

$$W_{ab} = 2\Phi\sqrt{\gamma_a \gamma_b} \tag{9.6}$$

where Φ is the ratio of molar volumes of materials at the interface, which is 0.626 for magnetite sliding on steel. The temperature dependence of the surface energy is:

$$\gamma(T) = \gamma_0 E(T)/E_0 \tag{9.7}$$

where T is temperature in °C, E is the Young modulus, and subscript 0 denotes the values at room temperature. At room temperature, steel and magnetite have a similar Young modulus and surface energy, 210 GPa and 1.5 J m^{-2}, respectively, whereas temperature dependence of the Young modulus (GPa) was given as:

$$E_{steel} = -0.1413T + 210.81 \qquad [9.8]$$

$$E_{Fe3O4} = -1.004 \times 10^{-7}T^3 + 2.26 \times 10^{-5}T^2 - 2.102 \times 10^{-2}T + 210.3 \qquad [9.9]$$

Interestingly, Jupp and Beynon (2005) successfully validated Straffelini's model, rather than their own, on a rig (Section 4.8).

References

Bisson, E.E. 1972. Boundary lubrication. In *Advanced Bearing Technology*, E.E. Bisson and W.J. Anderson (Eds.), 15–61. NASA SP-38.

Ferrante, J. 1988. *Applications of Surface Analysis and Surface Theory in Tribology*. NASA TM 101392.

Halling, J. 1982. The role of surface films in the frictional behaviour of lubricated and 'dry' contacts—A unifying influence in tribological theory. *ASLE Trans.* 25:528–536.

Jupp, S.P., and J. Beynon. 2005. A study of friction in non-lubricated high temperature steel processing. *Steel Res. Int.* 76:387–391.

Jupp, S.P., J. Talamantes-Silva, and J.H. Beynon. 2004. Application of fundamental friction model to the hot rolling of steel. In *Metal Forming 2004*, J. Kusiak, P. Hartley, and J. Majta et al. (Eds.), 325–329. Bad Harzburg. GRIPS Media.

Kragelsky, I.V., M.N. Dobychin, and V.S. Kombalov. 1977. *Fundamentals of the Calculations of Friction and Wear*. Moscow: Mashinostroenie.

Kragelsky, I.V. and V.S. Shchedrov. 1956. *Development of the Science of Friction: Dry Friction*. Moscow: Academy of Sciences of USSR.

Lancaster, J.K. 1963. The formation of surface films at the transition between mild and severe metallic wear. *Proc. R. Soc. A* 273:466–483.

Ludema, K.C. 1996. Mechanism-based modelling of friction and wear. *Wear* 200:1–7.

Pasaribu, H.R. 2005. *Friction and Wear of Zirconia and Alumina Ceramics Doped with CuO*. PhD diss., University of Twente, Enschede.

Straffelini, G. 2001. A simplified approach to the adhesive theory of friction. *Wear* 249:79–85.

10

Friction on Atomic and Molecular Scales

Understanding the fundamentals of friction needs studies at the atomic/molecular levels. This requires clean surfaces and vacuum to avoid contamination and oxidation, yet these nuisances are present in practical situations. On the other hand, it is the only way to understand the fundamental phenomena governing friction. In this section, issues specific to friction on atomic scale are presented first, followed by the basics of metallic friction.

10.1 Some Issues Specific to Atomic-Scale Friction

There are two major differences between the friction at macroscale and microscale

1. Burke (2003) noted that macroscopic friction is often interpreted according to Gyalog and Thomas (1997), as 'the force needed to plastically deform interlocking asperities of surfaces in relative motion'. However, she noted, friction exists between atomically flat surfaces. A possible solution is offered by Hölscher, Schirmeisen, and Schwarz (2008), in that stick-slip movement is the underlying mechanism for atomic-scale friction (Figure 10.1).
2. Krim (2002) argued that, theoretically, at atomic scale, there should be no static friction between clean surfaces. We observe it in real life at macroscale due to the presence of surface contaminants.

Krim (2002) noted that if commensurability changes to incommensurability, sliding friction is reduced drastically at atomic scale. This was explained via Figure 10.2 by Hölscher et al. (2008):

1. Atoms in the substrate form valleys and peaks that an incoming atom must overcome.
2. Interatomic distance in the substrate is a. Two atoms, spaced by b, slide over it. If $b \neq a$, they hold together and cannot fall as deeply into the valleys as a single atom.

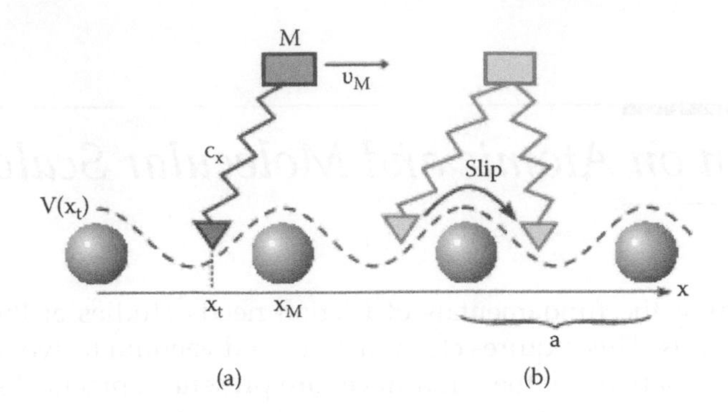

FIGURE 10.1
A point-like tip, coupled elastically to body M, slides on an atomically flat surface. At $x_1 = x_M$ the body is in equilibrium. In the valley, the body is 'stuck', but being pulled at the velocity v_M, it slips over the atom. This constitutes the stick-slip motion. (Reprinted from H. Hölscher, A. Schirmeisen and U.D. Schwarz, *Phil. Trans. A* 366:1383–1404, 2008. (With permission from the Royal Society.)

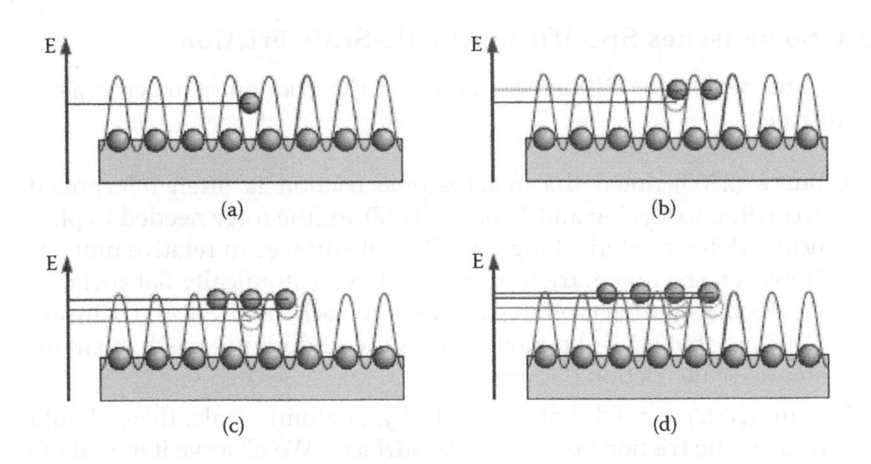

FIGURE 10.2
Incommensurability: although the number of atoms that have to overcome the barrier increases, the barrier height decreases significantly. For increasingly larger contacts, the effective overall barrier height approaches zero. (Reprinted from H. Hölscher, A. Schirmeisen and U.D. Schwarz, *Phil. Trans. A* 366:1383–1404, 2008. With permission from the Royal Society.)

3. As more atoms slide, the less they drop, and less force is needed to overcome the valleys. This effect is called superlubricity or structural lubricity*.

* A more rigorous definition was by Hirano (2003); superlubricity occurs when two conditions are met, namely each atom follows its equilibrium position adiabatically, and the contacting crystal surfaces are incommensurate.

10.2 Friction of Metals

On the macroscopic scale, frictional force between two sliding metals is the sum of shearing and ploughing forces (Bowden and Tabor, 1950). They contended that shearing is more important, which is supported by others (Bisson, 1968; Czichos, 1978; Schey, 1983)*. Rabinowicz and Tabor (1951) investigated the pick-up of metal by rubbing surfaces, and concluded that friction between metals is largely due to the formation and shearing of metallic junctions. Shearing is about breaking adhesion between surfaces, so the adhesive aspects of metals on the atomic scale are addressed. Three aspects of the propensity for adhesion of metals are considered, namely the mutual solubility of metal pairs, chemical activity on the surface of individual metals and the tensile/shear strength of a metal.

10.2.1 Solubility of Metal Pairs

Peterson and Johnson (1952) suggested that adhesion ('welding') is stronger between metals that are mutually soluble†. Bisson (1968) quoted the criterion for the minimum scoring of metal pairs proposed by Roach, Goodzeit and Hunnicutt: 'Two metals can slide on each other with relatively little scoring if both the following conditions are met: (1) the two metals are insoluble in each other and (2) at least one of the metals is from the B-subgroup of the periodic table'.

Of common metals, the B-subgroup includes Ti, V, Zr, Nb, Cr, Mo, W, Nb, Cu, Ag, Au and Zn, whereas Fe, Al, Mg, Co and Ni are outside it. Apparently, it was observed that there are 114 pairs of metals which seem to support the criterion and only 9 which do not. It is noteworthy, though, that Johnson and Keller (1966) argued that the mutual solubility is not a sole criterion for adhesion, inasmuch as '... the condition for immiscibility involves a bulk material energy criterion, whereas the condition for adhesion requires a criterion involving surface and interfacial energies.'

10.2.2 Chemical Activity on Metal Surface

Kragelsky and Shchedrov (1956) considered that a measure of the chemical activity of a metal is the heat of formation of its oxides. If metal is less active, the force of its molecular interaction with another surface will be smaller. This explained the experimental results that the COF of a metal decreases with the increasing heat of formation of its oxides.

* According to Schey (1983), ploughing is important only if a very hard material slides over a soft solid. Czichos (1978) believes that in practical situations, COF due to ploughing does not exceed 0.05, although additional force is required to push the ploughed material, so this is the lower limit.

† Mutual solubility of metal pairs is determined by Hume–Rothery rules (Cottrell, 1967).

Buckley (1971) experimented in the argon atmosphere with the binary noble metal alloys sliding on iron, and observed that the greater the free energy of formation of the binary alloy, the lower the friction and wear when sliding on iron. This can be explained as follows:

1. The free energy of formation is a measure of the binding energy between metals in an alloy.
2. The binding energy depends on the valence electron interaction between these metals.
3. With stronger interaction between these metals in the alloy, there will be fewer valency electrons available at the interface for the interaction with iron.

Buckley (1976) reported another set of experiments that confirmed that the COF of individual metals decreases with chemical reactivity. As shown in Figure 10.3, the COF decreases with the increasing portion of the d-character of metallic bonding. Miyoshi and Buckley (1981a) argued that the filling of the d-valence electron band in transition metals determines its chemical stability. The activity of the surface of a metal decreases with the percentage of its d-bond character.

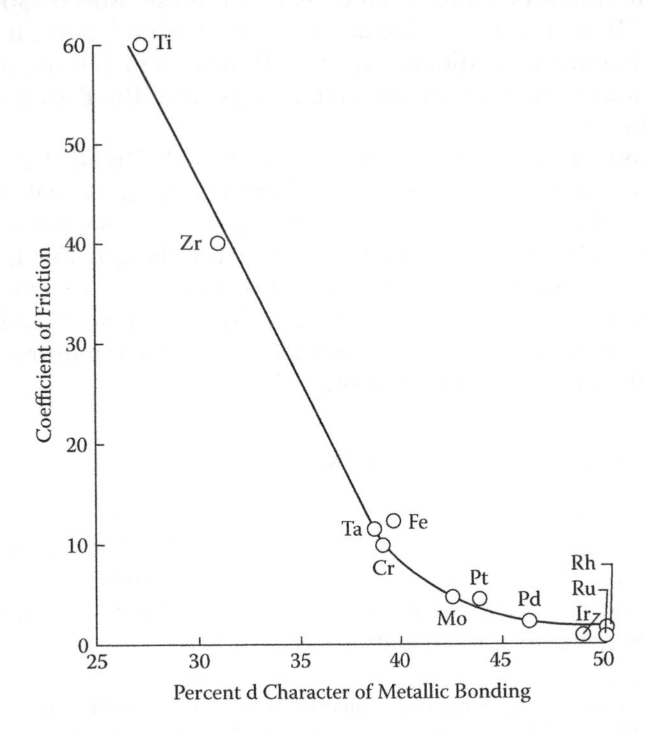

FIGURE 10.3
COF as a function of percent d-bond character for metals in contact with themselves. (Reprinted from D.H. Buckley, *Adhesion and Friction of Thin Metal Films*, 1976, NASA TN D-8230.)

10.2.3 Tensile and Shear Strength

Miyoshi and Buckley (1981a,b) observed that the COF decreases with ideal tensile or shear strength (Figure 10.4). The explanation was based on the premise that the filling of the d-valence electron band in transition metals is responsible for properties including cohesive energy and shear modulus:

1. As strong metallic or non-metallic material slides over metal, tearing and shearing occur over the contact surface.
2. The tearing and shearing may occur at the asperity junctions formed at the interface, or within the metal itself.
3. If the metal has low strength, it may be weaker than the asperity bonds at the interface.
4. It was observed that in the case of a weaker metal, much more of it is transferred to the slider than with stronger metals, and friction is higher. This can also be explained by the increased ploughing of slider over weaker metals.

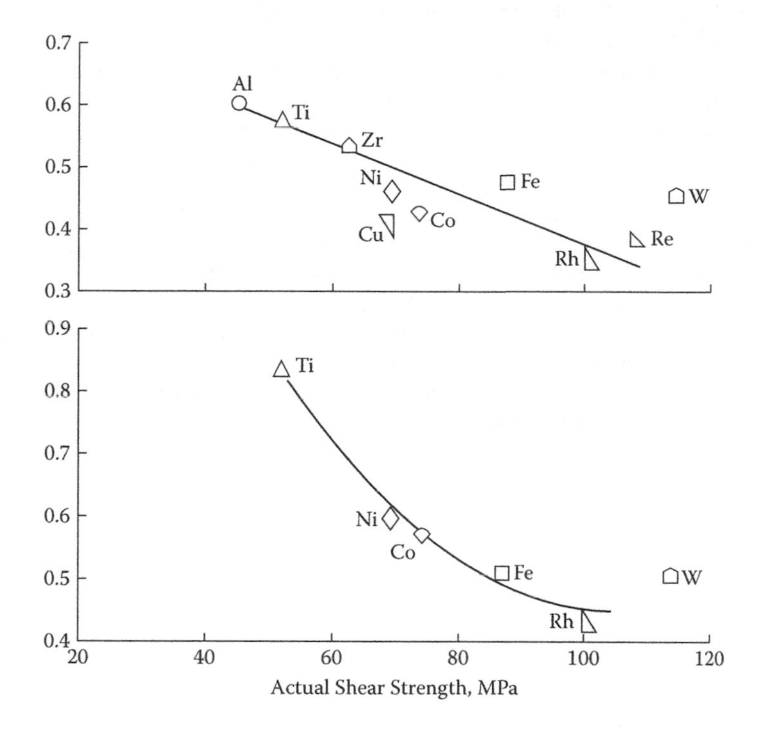

FIGURE 10.4
Effect of shear strength on coefficient of friction in sliding of SiC (top) and Mn–Zn ferrite (bottom), over various metals. (Reprinted from K., Miyoshi and D.H. Buckley, *Correlation of Ideal and Actual Shear Strengths of Metals with Their Friction Properties*, 1981b, NASA TP 1891.)

References

Bisson, E.E. 1968. *Friction, Wear and the Influence of Surfaces*. NASA TM/X-52380.
Bowden, F.P., and D. Tabor. 1950. *The Friction and Lubrication of Solids*. Oxford: Oxford University Press.
Buckley, D.H. 1971. *Thermochemistry of Binary Alloys and Its Effect upon Friction and Wear*. NASA TN D-6311.
Buckley, D.H. 1976. *Adhesion and Friction of Thin Metal Films*. NASA TN D-8230.
Burke, S.A. 2003. *Friction Force Microscopy: Seeking New Understanding of Friction from a Nanoscale Perspective*. Unpublished report. McGill University, 28 February.
Cottrell, A.H. 1967. *An Introduction to Metallurgy*. London: Edward Arnold.
Czichos, H. 1978. *Tribology—A System Approach to the Science and Technology of Friction, Lubrication and Wear*. Amsterdam: Elsevier.
Gyalog, T., and H. Thomas. 1997. Atomic friction. *Zeit. Physik B* 104:669–674.
Hirano, M. 2003. Superlubricity: A state of vanishing friction. *Wear* 254:932–940.
Hölscher, H., A. Schirmeisen, and U.D. Schwarz. 2008. Principles of atomic friction: From sticking atoms to superlubric sliding. *Phil. Trans. A* 366:1383–1404.
Johnson, K.I., and D.V. Keller. 1966. *The Effect of Contamination on the Adhesion of Metallic Couples in Ultra-High Vacuum*. NASA CR 71147.
Kragelsky, I.V., and V.S. Shchedrov. 1956. *Development of the Science of Friction: Dry Friction*. Moscow: Academy of Sciences of USSR.
Krim, J. 2002. Surface science and the atomic-scale origins of friction: What once was old is new again. *Surf. Sci.* 500:741–758.
Miyoshi, K., and D.H. Buckley. 1981a. *Relationship Between the Ideal Tensile Strength and the Friction Properties of Metals in Contact with Nonmetals and Themselves*. NASA TP 1883.
Miyoshi, K., and D.H. Buckley. 1981b. *Correlation of Ideal and Actual Shear Strengths of Metals with Their Friction Properties*. NASA TP 1891.
Peterson, M.B., and R.L. Johnson. 1952. *Friction and Surface Damage of Several Corrosion-Resistant Materials*. Washington: NACA, RM E51L20.
Rabinowicz, E., and D. Tabor. 1951. Metallic transfer between sliding metals: An autoradiographic study. *Proc. R. Soc. Lond. A* 208:455–475.
Schey, J.A. 1983. *Tribology in Metalworking. Friction, Lubrication and Wear*. Ohio: American Society for Metals.

11

Tribological Properties of Oxidised Metals and Carbides

On Earth, most surfaces are covered with oxide films, which hinder adhesion and reduce friction. Even at low concentrations, oxygen is quickly adsorbed and reduces friction considerably, as shown in Figure 11.1. In the vacuum of outer space, after oxides are removed from rubbing surfaces, they cannot re-form due to the lack of oxygen, and galling occurs (Fusaro, 1991). Also, although rolls in a hot strip mill consist mainly of iron, even before oxides are formed later in the rolling schedule, friction is much lower than that of pure iron due to the adsorption of atmospheric oxygen.

This section deals with a general link between oxides and friction, however, frictional properties of iron oxides are tackled in detail, given the importance of steel as an engineering material. This book pays particular attention to the hot rolling of steel, hence chromia, molybdenum oxide and oxide glaze are also analysed, inasmuch as they may form on the roll surface.

Although carbides are neither lubricants nor have low friction, they influence friction, as shown in Chapter 16. NASA conducted extensive experiments which showed that their mixing with solid lubricants slows down their wear and lowers friction. The properties of key carbides are discussed, with an emphasis on chromium carbides, which have a significant presence in the rolls in early stands of a hot strip mill.

11.1 General

11.1.1 Tribological Properties of Oxides

Kragelsky and Shchedrov (1956) observed that COF may increase or decrease with oxide thickness. They contended that thin oxide attenuates the forces of molecular interaction. However, with thicker layers the volumetric properties of oxide, such as structure, tend to dominate the friction instead of film thickness. This is consistent with the observations by Hirst and Lancaster (1954):

1. For some metals the first Amontons' law does not hold above certain load.
2. Above that load, oxide film breaks and metals come into direct contact.

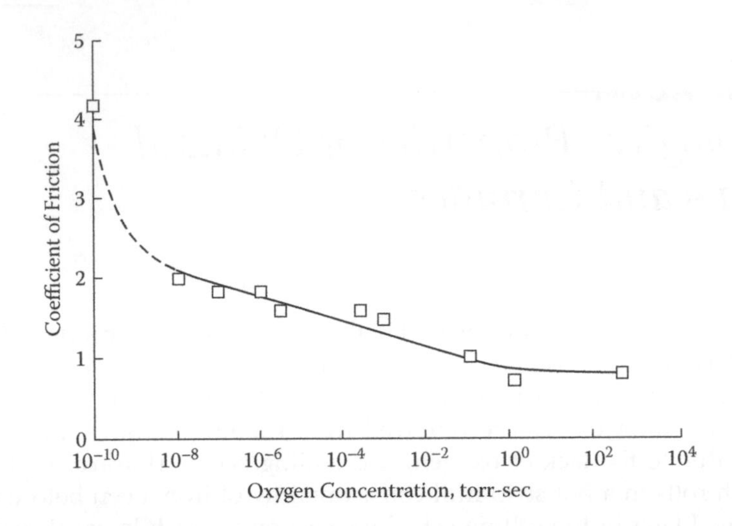

FIGURE 11.1
Effect of oxygen concentration on COF for iron. (Reprinted from D.H. Buckley, *Influence of Chemisorbed Films on Adhesion and Friction of Clean Iron*, 1968, NASA TN D-4775.)

3. For an individual metal, this critical load mainly depends on the rate of formation of surface oxide film, rather than on its thickness.
4. The film formed by slow oxidation possesses 'the original topographical features of the underlying metal'. In rapid oxidation, excrescences of oxide stick out and break easily. A rapidly formed oxide film can completely negate its lubricity.
5. Oxide reduces friction when it has similar physical properties as its metal, and the combinations of hard oxide and soft metal are to be avoided (Section 7.3).

Peterson et al. (1990) identified four regimes of oxide formation and wear:

1. Fast oxidation, fast wear. As soon as it is formed, oxide is removed. Too little of it forms at a given wear rate, or it detaches easily. The controlling factor is the oxidation rate.
2. Slow oxidation, fast wear. Oxide grows slowly and detaches after reaching a critical thickness. Controlling factors are the rate of removal and the critical thickness.
3. Fast oxidation, slow wear. It is claimed that this regime is not observed.
4. Slow oxidation, slow wear, which results in a stable film. The rate of wear is the function of film thickness. If the film is thick, the rate of wear increases; if it is too thin, oxidation increases. This agrees with the observation by Hirst and Lancaster (1954) that the highest load capacity is on the films formed slowly at low temperatures.

Peterson and co-workers also defined the list of properties of effective oxide films:

1. Film thickness is 0.05–1 μm (although oxide on rolls in hot rolling is much thicker).
2. Oxide is soft and ductile, and not abrasive.
3. Wear rate is lower than the oxidation rate.
4. Film fails by wear, not fatigue or fracture.
5. Slow oxidation, slow wear applies.
6. Metal and oxide have similar coefficients of thermal expansion.
7. Oxide can be efficiently compacted.
8. There is slip at the metal/oxide interface.
9. Oxide forms low-melting-point glassy oxide.

Erdemir (2000) offered a comprehensive crystal-chemical model capable of describing the shear rheology, or lubricity of certain oxides at high temperatures. Unfortunately, only FeO is included of the oxides of interest to hot rolling of steel.

11.1.2 Strength of Metal–Oxide Interface

Experiments suggest that, in clean systems, the metal–oxide interfaces are generally strong, and failure occurs in oxide or metal (Evans, Hutchinson, and Wei 1999). The interface can be embrittled and weakened by segregrants and contaminants, particularly in the presence of moisture. Interestingly, Sullivan (1987) claimed that '...the adhesive forces between oxide and metal are usually lower than the cohesive forces in either oxide or metal'. The difference of views can be explained by the cleanliness of specimens. For example, Hou (2005) reported that small amounts of sulphur weaken the metal–oxide interface. On the other hand, reactive elements such as Hf, Y and Zr react with sulphur and tie it up, decreasing its negative impact on oxide adhesion. This agrees with the observations by Finnis (1996) on the role of Y, and by Lees (2003) regarding Ce, Zr and Y.

11.2 Iron Oxides

Magnetite (Fe_3O_4), wüstite ($Fe_{0.947}O$) and hæmatite (Fe_2O_3) play key roles in determining the friction in the roll gap. Good lubricity of wüstite and magnetite was noted by Johnson and colleagues (1947 and 1948, respectively). The latter also observed the poor lubricity of hæmatite. Luong and Heijkoop

TABLE 11.1
Oxide Composition in Various Tests

Mixture	Oxide Thickness (μm)	FeO (wt-%)	Fe_3O_4 (wt-%)	Fe_2O_3 (wt-%)
A	150	100	0	0
B	220	100	0	0
C	500	100	0	0
D	230	59	33	8
E	200	23	52	25

Source: Tests carried out by Luong and Heijkoop (1981). Reprinted with permission from Elsevier.

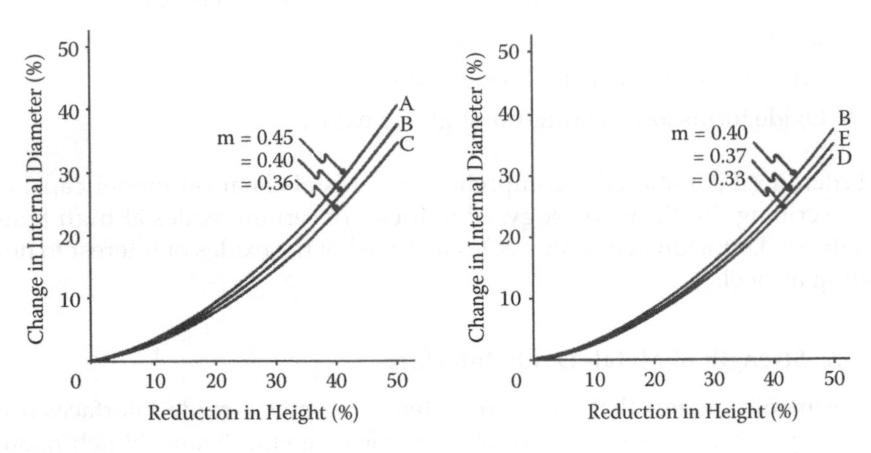

FIGURE 11.2
The value of the friction factor for mixtures of iron oxides in Table 11.1. Note the use of the friction factor (Section 17.2.2). (Reprinted from L.H.S. Luong, and T. Heijkoop, *Wear* 71:93–102, 1981. With permission from Elsevier.)

(1981) conducted systematic tests (see Table 11.1) to establish the impact of these iron oxides on friction. In tests with pure wüstite, the COF decreased with increasing oxide thickness (Figure 11.2). Regarding the mixtures of all three iron oxides, friction was lower in the test D, with smaller hæmatite content than in test E (oxide was thicker in test D, albeit slightly).

Several other observations and comments are of interest:

1. Bailey and Godfrey (1954) observed that fretting of magnetite produced hæmatite, which coincided with rising friction. COF remained low when magnetite did not fret.
2. Lu, Cotter, and Eadie (2005) observed that COF increases with the hæmatite content in the film separating the rubbing disks.
3. Dmitriev et al. (2008) modelled friction in a brake, where both the pad and the disc were ferrous. COF decreased from 0.7–0.9 to 0.4 after they were coated with magnetite.

4. Joos et al. (2007) and Vergne et al. (2006a) claimed that thicker scale reduces friction, which is consistent with lower shear strength; soft oxides (presumably FeO and Fe_3O_4) are lubricants, whereas hard oxides (presumably Fe_2O_3) are abrasives.
5. Vergne et al. (2006a) stated, without elaboration, that adherent oxide increases friction regardless of its hardness. If it is loose, oxide decreases friction only if it is soft.

Little elaboration is given in the literature as to the better lubricity of wüstite and magnetite. According to Johnson, Godfrey, and Bisson (1948) 'Small crystals [of magnetite] growing on a steel surface adapt themselves to the irregularities of the steel more readily than larger ones [of hæmatite] and therefore make the film as a whole more tenacious'. A possible explanation is the shear strength difference. As shown in Chapter 4.4, low shear strength reduces friction. Hæmatite has higher shear strength than iron, that is, 167 kg mm^{-2} versus 100 kg mm^{-2} according to Buckley (1971). Also, Jarl (1993) indicated that wüstite is softer than iron, whereas Lundberg and Gustaffson (1994) implied that wüstite and magnetite are softer than hæmatite (see detailed oxide properties in Appendix D).

11.3 Chromium and Molybdenum Oxides

There are two reasons why Cr_2O_3 (chromia) is of interest. First, Vergne et al. (2006b) contended that it constitutes most of the oxide on HiCr rolls, and some on HSS rolls. Second, as part of an investigation into solid lubricants for elevated temperatures, NASA conducted significant research into its tribology:

1. Pomey (1952) reported that chromia adheres well and provides good (though unquantified) lubrication on steels with composition similar to that of HiCr rolls.
2. According to DellaCorte (1987a), in lightly loaded, conforming contacts such as bearings, this oxide forms a thin coherent film that is a good lubricant. However, in high loads at non-conforming contacts, such as a pin on a disk, this oxide acts as an abrasive.
3. DellaCorte and Sliney (1987) found that chromia which forms on the plasma-sprayed coating acts as an abrasive in the sliding contact, increasing friction and wear. However, it is a good antiwear coating when applied as a smooth adherent coating.

NASA researchers mixed chromia with primary lubricants to enhance their wear resistance. Initial coatings had a low coefficient of thermal expansion

(DellaCorte and Fellenstein, 1996), leading to cracking and spalling (Striebing et al., 2007). Eventually, a mixture with similar thermal expansions of coating and substrate was devised (DellaCorte, 1998). It contained 60 wt-% NiCr as a binder, 20 wt-% Cr_2O_3 as a hardening additive and 10 wt-% each of Ag and BaF_2/CaF_2 eutectic, as low- and high-temperature lubricants, respectively.

Molybdenum oxides MoO_2 and MoO_3 are of interest given the high Mo content in HSS rolls. MoO_2 is very abrasive (Ahn, Lyo, and Lim 2000), and the findings for MoO_3 are inconclusive. According to Bisson et al. (1956), it is abrasive. However, several authors considered it a lubricant (Sliney and Bisson, 1964; Peterson, Li, and Murray 1993; Ahn et al. 2000).

11.4 Oxide Glaze

This glaze is formed between two sliding metallic surfaces, from the debris generated by wear that may be subsequently oxidised and sintered (Inman, Rose, and Datta 2006a). Bailey and Godfrey (1954) noticed that oxide particles could be pressed into a film. This was somewhat surprising, given that oxides are hard and lack plasticity. This film can reduce friction and wear, but may also be abrasive, depending on the chemistry of oxides. Also, as shown above, Peterson et al. (1990) considered that oxide film is good for wear protection if it forms a glassy oxide. The rate of formation is influenced by speed, temperature and load.

Stott and Jordan (2001) experimented with HiCr and HSS pins on C-steel disks, and observed that loose debris cannot be retained between surfaces if the debris is large, or wear grooves are shallow. The size of wear increases with load, and grooves are shallow on harder surfaces. They also observed that a more protective glaze was formed on HiCr substrate. Stott (2002) attributed the formation of oxide to three mechanisms:

1. Broken metallic debris oxidises very quickly; even at room temperature, a 2-nm-thick oxide layer may form in less than 0.1 s.
2. Metal oxidises at the contacting asperities, as well as in the non-contact area. This oxide is eventually removed, particularly at high speed, generating more oxide debris, and exposing the bare metal surface, which is oxidised again.
3. At higher temperatures, a tenacious oxide layer may be formed, protecting the substrate.

Regarding the behaviour of debris, three scenarios are possible, namely its removal, retention between the contacting surfaces where it acts as an abradent and its adherence to one or both surfaces, protecting them against further wear.

Systematic tests with Stellite 6 and Nimonic 80A showed a very complicated impact of speed and temperature on oxide formation and wear, and the results regarding friction are still inconclusive (Inman et al., 2006a,b; Inman and Datta, 2008). Also, glaze was found to decrease friction in hot rolling laboratory tests (Milan et al., 2005; Pellizzari, Cescato, and de Flora, 2009).

In railway engineering, it was observed that debris is generated by wear at the wheel–rail interface (Lu et al., 2005), and oxidised into magnetite and hæmatite. This oxide forms a low-friction slurry with water, but when it is dry it forms high-friction layers. Generally, solid debris, residual lubricants and contaminants (sand, leaves, dust) form the so-called 'third body', the friction of which is strongly related to rheological properties. Quinn (1983) proposed two mechanisms for debris formation that may result in the oxide glaze, and favoured the second one:

1. The bulk of oxidation occurs at the instant the virgin metal is exposed, and further contacts shear the oxide at the metal–oxide interface.
2. Equal oxidation occurs at each contact until a critical oxide thickness is reached, beyond which shearing occurs at the metal–oxide interface.

11.5 Properties of Key Carbides

The metallic carbides of interest are those found in roll shells, with the following general properties (Kieffer and Benesovsky, 1988):

1. *Interstitial carbides,* formed with metals from groups 4–6 in the periodic table, with the exception of Cr. These metals have a relatively large atomic radius, and include V, W and Mo, which are present in HSS and HiCr rolls. These carbides are chemically inert, very hard and retain strength at elevated temperatures;
2. *Intermediate carbides,* with the metals with smaller atomic radius, such as Fe and Co. They are more reactive than interstitial carbides and have a more complex structure.
3. *Chromium carbides* have properties somewhere between those types above.

Each of the mentioned metals has a variety of carbides of different structure (Table 11.2). Metallic carbides are generally represented with the formula M_xC_y, where M represents the total metal atoms. For example, Fe_4W_2C

TABLE 11.2
Carbides of Practical Interest

Metal	Carbide	Structure	Reference
Chromium	Cr_3C	Orthorhombic	Martin, 2006
	Cr_3C_2	Orthorhombic	Warlimont, 2005
	Cr_7C_3	Hexagonal	Warlimont, 2005
	$Cr_{23}C_6$	Face-centred cubic	Goodwin et al., 2005
Molybdenum	MoC	Hexagonal	Warlimont, 2005
	Mo_2C	Hexagonal	Warlimont, 2005
	Mo_3C_2	Hexagonal	Pearson, 1972
Tungsten	WC	Hexagonal	Lassner and Schubert, 1999
	W_2C	Hexagonal	Goodwin et al., 2005
Vanadium	VC	Face-centred cubic	Goodwin et al., 2005
	V_2C	Orthorhombic at <800°C	Toth, 1971
	V_4C_3	Rock salt	Frad, 1968
Iron	Fe_2C	Hexagonal	Frad, 1968
	Fe_3C	Orthorhombic	Krauss, 2005
	Fe_7C_3	Orthorhombic or hexagonal	Fang et al., 2009

is an M_6C carbide. Carbides like this one, containing two metals, are double carbides. The hardness of some carbides is listed in Appendix D.

Carbide Cr_3C_2 was extensively trialled to improve the wear resistance of high-temperature solid lubricants. It was bonded with Ni, Al and Co, and mixed in various ratios with solid lubricants Ag and BaF_2/CaF_2 (DellaCorte, 1987b; see also Figure 11.3). It had a good wear resistance and thermal stability, but high friction on its own (Brainard, 1977; DellaCorte 1987a). However, without its addition, the solid lubricants were quickly removed from the interface, and friction increased sharply. DellaCorte (1987b) tested various combinations of bonded carbide and lubricants. When the carbide content

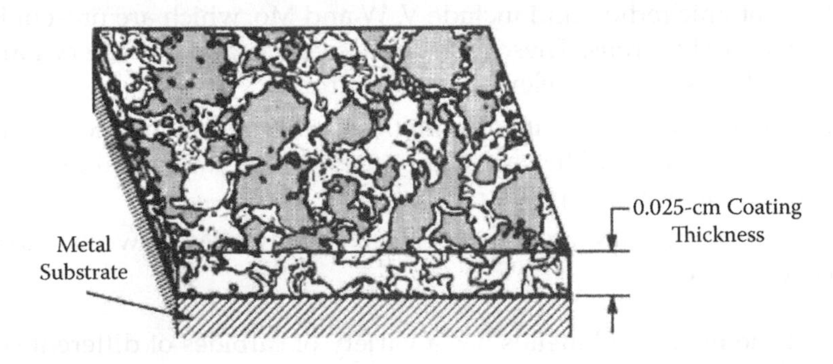

FIGURE 11.3
Microstructure of mixture PS200 with 80 wt-% bonded carbide and 10 wt-% of each eutectic. (Reprinted from H.E. Sliney, *Aeropropulsion '87*, 1987, 89–101.)

became too low for the formation of a continuous network, lubricant was ploughed, and friction and wear increased. Microstructure of those mixtures is similar to the microstructure of HiCr rolls in Chapter 15.

References

Ahn, H.-S., I.-W. Lyo, and D.S. Lim. 2000. Influence of molybdenum composition in chromium oxide-based coatings on their tribological behaviour. *Surf. Coating Tech.* 133–134:351–361.

Bailey, J.M., and D. Godfrey. 1954. *Coefficient of Friction and Damage to Contact Area During the Early Stages of Fretting. iii - Steel, Iron, Iron Oxide, and Glass Combinations.* Washington: NACA TN 3144.

Bisson, E.E., R.L. Johnson, M.A. Swikert et al. 1956. *Friction, Wear and Surface Damage of Metals as Affected by Solid Surface Films.* Washington: NACA Report 1254.

Brainard, W.A. 1977. *Friction and Wear Properties of Three Hard Refractory Coatings Applied by Radiofrequency Sputtering.* NASA TN D-8484.

Buckley, D.H. 1968. *Influence of Chemisorbed Films on Adhesion and Friction of Clean Iron.* NASA TN D-4775.

Buckley, D.H. 1971. *Thermochemistry of Binary Alloys and Its Effect upon Friction and Wear.* NASA TN D-6311.

DellaCorte, C. 1987a. *Composition Optimization of Chromium Carbide Based Solid Lubricant Coatings for Foil Gas Bearings at Temperatures to 650°C.* NASA CR 179649.

DellaCorte, C. 1987b. *Experimental Evaluation of Chromium-Carbide-Based Solid Lubricant Coating for Use to 760°C.* NASA CR 180808.

DellaCorte, C. 1998. *The Evaluation of a Modified Chrome Oxide Based High Temperature Solid Lubricant Coating for Foil Gas Bearings.* NASA TM-208660.

DellaCorte, C., and H.E. Sliney. 1987. *Effects of Atmosphere on the Tribological Properties of a Chromium Carbide Based Coating for Use to 760°C.* NASA TM-88894.

DellaCorte, C., and J.A. Fellenstein. 1996. *The Effect of Compositional Tailoring on the Thermal Expansion and Tribological Properties of PS300: A Solid Lubricant Composite Coating.* NASA TM 107332.

Dmitriev, A.I., A.Yu. Smolin, S.G. Psakhie et al. 2008. Computer modelling of local tribological contacts by the example of the automotive brake friction pair. *Phys. Mesomechanics* 11:73–84.

Erdemir, A. 2000. A crystal-chemical approach to lubrication by solid oxides. *Trib. Lett.* 8:9–102.

Evans, A.G., J.W. Hutchinson, and Y. Wei. 1999. Interface adhesion: Effects of plasticity and segregation. *Acta Mater.* 47:4093–4113.

Fang, C.M., M.A. van Huis, and H.W. Zandbergen. 2009. Structural, electronic and magnetic properties of iron carbide Fe_7C_3 phases from first-principles theory. *Phys. Rev. B* 80:1–9.

Finnis, M.W. 1996. The theory of metal-ceramic interfaces. *J. Phys. Condens. Matter* 8:5811–5836.

Frad, W.A. 1968. Metal carbides. In *Advances in Inorganic Chemistry and Radiochemistry, Vol. 11*, H.J. Emeléus and A.G. Sharpe (Eds.), 153–237. New York: Academic Press.

Fusaro, R.L. 1991. *Tribology Needs for Future Space and Aeronautical Systems*. NASA TM 104525.

Goodwin, F., S. Guruswamy, K.U. Kainer et al. 2005. Metals. In *Springer Handbook of Condensed Matter and Materials Data*, W. Martienssen and H. Warlimont (Eds.), 431-476. Berlin: Springer.

Hirst, W., and J.K. Lancaster. 1954. The influence of oxide and lubricant films on the friction and surface damage of metals. *Proc. R. Soc. Lond. A* 223:324–338.

Hou, P.Y. 2005. *Segregation Phenomena at Growing Alumina/Alloy Interfaces*. Lawrence Berkeley National Laboratory: Rep. LBNL-52959.

Inman, I.A., and P.S. Datta. 2008. Development of a simple 'temperature versus sliding speed' wear map for the sliding wear behaviour of dissimilar metallic interfaces II. *Wear* 265:1592–1605.

Inman, I.A., R.S. Rose, and P.K. Datta. 2006a. Studies of high-temperature sliding wear of metallic dissimilar interfaces II: Incoloy MA956 versus Stellite 6. *Trib. Int.* 39:1361–1375.

Inman, I.A., S.R. Rose, and P.K. Datta. 2006b. Development of a simple 'temperature versus sliding speed' wear map for the sliding wear behaviour of dissimilar metallic interfaces. *Wear* 260:919–932.

Jarl, M. 1993. An estimation of the mechanical properties of the scale at hot rolling of steel. In *1st Int. Conf. Model. Metal Rolling Proc., 21–23 Sep. 1993, London*, 614–628.

Johnson, R.L., M.A. Swikert, and E.E. Bisson. 1947. *Friction at High Sliding Velocities*. Washington: NACA TN 1442.

Johnson, R.L., D. Godfrey, and E.E. Bisson. 1948. *Friction of Solid Films on Steel at High Sliding Velocities*. Washington: NACA TN 1578.

Joos, O., C. Boher, C. Vergne et al. 2007. Assessment of oxide scales influence on wear damage of HSM work rolls. *Wear* 263:198–206.

Kieffer, R., and F. Benesovsky. 1978. Carbides (industrial heavy-metal). In *Kirk-Othmer Encyclopaedia of Chemical Technology Vol. 4*, 476–535. New York: John Wiley & Sons.

Kragelsky, I.V., and V.S. Shchedrov. 1956. *Development of the Science of Friction: Dry Friction*. Moscow: Academy of Sciences of USSR.

Krauss, G. 2005. *Steels: Processing, Structure and Performance*. Materials Park: ASM International.

Lassner, E., and W.-D. Schubert. 1999. *Tungsten: Properties, Chemistry, Technology of the Elements, Alloys, and Chemical Compounds*. Berlin: Springer.

Lees, D.G. 2003. The influence of sulphur on the adhesion and growth mechanisms of chromia and alumina scales formed at high temperatures: The sulphur effect. *Proc. R. Soc. Lond. A* 459:1459–1466.

Lu, X., J. Cotter, and D.T. Eadie. 2005. Laboratory study of the tribological properties of friction modifier thin films for friction control at the wheel/rail interface. *Wear* 259:1262–1269.

Lundberg, S.-E., and T. Gustaffson. 1994. The influence of rolling temperature on roll wear, investigated in a new high temperature test rig. *J. Mat. Proc. Tech.* 42:239–291.

Luong, L.H.S., and T. Heijkoop. 1981. The influence of scale on friction in hot metal working. *Wear* 71:93–102.

Martin, J.W. (ed.). 2006. *Concise Encyclopedia of the Structure of Materials*. Oxford: Elsevier.

Milan, J.C.G., M.A. Carvalho, R.R. Xavier et al. 2005. Effect of temperature, normal load and pre-oxidation on the sliding wear of multi-component ferrous alloys. *Wear* 259:412–423.

Pearson, W.B. 1972. *The Crystal Chemistry and Physics of Metals and Alloys*. New York: John Wiley & Sons.

Pellizzari, M., D. Cescato, and M.G. de Flora. 2009. Hot friction and wear behaviour of high speed steel and high chromium iron for rolls. *Wear* 267:467–475.

Peterson, M.B., S.J. Calabrese, S.Z. Li et al. 1990. Frictional properties of lubricating oxide coatings. *Tribol. Ser.* 17:15–25.

Peterson, M.B., S.Z. Li, and S.F. Murray. 1993. *Wear-Resisting Oxide Films for 900°C*. Argonne National Lab: Contract Number 20082401.

Pomey, J. 1952. *Friction and Wear*. Washington: NACA TM 1318.

Quinn, T.F.J. 1983. *NASA Interdisciplinary Collaboration in Tribology. A Review of Oxidational Wear*. NASA CR 3686.

Sliney, H.E. 1987. Self-lubricating coatings for high-temperature applications. In *Aeropropulsion '87*, 89–101.

Sliney, H.E., and E.E. Bisson. 1964. *Solid Lubricants for High Temperature Metal Working*. NASA Report TM-X-57741.

Stott, F.H. 2002. High-temperature sliding wear of metals. *Trib. Int.* 35:489–495.

Stott, F.H., and M.P. Jordan. 2001. The effects of load and substrate hardness on the development and maintenance of wear-protective layers during sliding at elevated temperatures. *Wear* 250:391–400.

Striebing, D.R., M.K. Stanford, C. DellaCorte et al. 2007. *Tribological Performance of PM300 Solid Lubricant Bushings for High Temperature Applications*. NASA/TM-2007-214819.

Sullivan, J.L. 1987. The role of oxides in the protection of tribological surfaces. II. In *Proc. Tribology—Friction, Lubrication and Wear. Fifty Years on, Vol. 1*, 293–301.

Toth, L. (ed.). 1971. *Transition Metal Carbides and Nitrates*. New York: Academic Press.

Vergne, C., C. Boher, R. Gras et al. 2006a. Influence of oxides on friction in hot rolling: Experimental investigations and tribological modelling. *Wear* 260:957–975.

Vergne, C., D. Batazzi, C. Gaspard et al. 2006b. Contribution of laboratory tribological investigations on the performance appraisal of work rolls for hot strip mill. In *Proc. ATS Rolling Conf., Paris, June 2006*.

Warlimont, H. 2005. Ceramics. In *Springer Handbook of Condensed Matter and Materials Data*, W. Martienssen and H. Warlimont (Eds.), 431–476. Berlin: Springer.

Section III

From Theoretical Concepts to Industrial Hot Rolling Processes

12

Chemical Composition and Microstructure of the Shells of HSS, HiCr and ICDP Work Rolls

Happy he who hath availed to know the causes of things.

Virgil, 70 BC–19 BC

A saying is attributed to Wolfgang Pauli: 'The solids were created by God, but surfaces were the work of the Devil', and it is on the surfaces that friction occurs. Atoms or molecules inside a solid interact with similar atoms or molecules. Those on the surface interact with those atoms or molecule inside the solid, but even more with the totally different substances and phenomena in the outside world.

What can be said at all can be said clearly.

Wittgenstein, 1922

Like a circle, like a ring
There is order in all things

Crime and the City Solution, 1990

Friction of HSS and HiCr rolls is linked to the chemical composition of their shells (see Chapter 16). Hence, after explaining the basic terminology, the composition of the shells of common roll types is shown. Structure and content of carbides are analysed next, followed by the roles of key chemical elements. The impact of roll microstructure on friction is also addressed, as well as the impact of a novel research area, the addition of rare earths.

Three basic types of rolls used in analyses are as follows:

1. *High-Speed Steel:* Named after the ability of high-speed steel tools to operate at cutting speeds well above those with common steel tools (Higgins, 1983). HSS retains hardness at high temperatures, and resists tempering by the heat generated at high speeds.
2. *High-Chromium:* It is iron or steel with high chromium content.

3. *ICDP (Indefinite Chilled—Double Poured)*: According to Schröder (2003) they are 'indefinite' because their hardness changes over radius '... is continuous but cannot be clearly defined'. Regarding the double pouring, the basic issue is how to produce roll with two distinctly different materials, shell and core. In double pouring, one 'material is cast into a chill mould and after solidification the other material is cast'.

Primary carbides are formed during solidification, and *secondary carbides* during heat treatment (Collins, 2002). *Eutectic carbide* is also formed during solidification as one of the mutually insoluble phases in ferrous alloys*. Regarding common carbides in the rolls, Lecomte-Beckers et al. (2004) consider MC, M_2C, M_6C and M_7C_3 eutectic, unlike $M_{23}C_6$.

12.1 Elemental Composition of Rolls and Carbides

HiCr and HSS rolls consist of a matrix with embedded carbides. The matrix is commonly tempered martensite, although bainite may be present (Molinari et al., 2000; Sorano, Oda, and Zuccarelli, 2004). Graphite is usually absent (Hashimoto et al., 1995; Sorano et al., 2004). The elemental composition of rolls in the commercial five-stand mill (from which the data were extracted for analyses in Chapters 15–17) and in the literature is in Tables 12.1–12.3.

12.2 Carbide Structure and Content

MC carbides have granular structure, M_6C are rod-like and M_3C and M_7C_3 form networks (Hashimoto et al., 1995), and Breyer, Skoczynski, and Walmag (1997) described M_2C as rod-like. Networked carbides have Cr as the dominant nonferrous component, probably due to the propensity of Cr to build complex carbides. The morphology of carbides is nicely illustrated in Figure 12.1, and the basic types of carbides in commercial rolls can be seen in Figure 12.2.

Given the high content of Cr and its carbides, the microstructure of HiCr rolls differs from HSS rolls (*cf.* Figures 12.3 and 15.7). HSS rolls contain mainly MC, M_2C, M_6C and M_7C_3 carbides, with some M_4C_3, M_8C_7 and

* Eutectic alloy is one where all components solidify or melt at the same temperature (Campbell, 2008).

TABLE 12.1

Elemental Composition of HSS Rolls (in wt-%)

	Mill Rolls	Literature	References
C	1.7–3.4	0.8–3.5	Sano et al., 1992; Collins, 2002; Kim et al., 2004
Ni		0–1.7	Hashimoto et al., 1995; Collins, 2002; Kim et al., 2004; Belzunce et al., 2004
Cr	1.9–10	2–13	Sano et al., 1992; Hashimoto et al., 1995; Collins, 2002
Mo	0.4–8	0.3–10	Hashimoto et al., 1995; Collins, 2002; Sorano et al., 2004
V	6–7	1–15	Goto et al., 1992; Sano et al., 1992; Hashimoto et al., 1995; Collins, 2002
W		0.5–20	Sano et al., 1992; Lecomte-Beckers et al., 1997; Collins, 2002; Garza-Montes-de-Oca et al., 2011
Si	0.5–1.2	0.2–1	Lecomte-Beckers et al., 1997; Park et al., 1999; Kim et al., 2004
Mn	0.1–1.1	0.3–0.9	Park et al., 1999; Kang et al., 2001; Lee et al., 2001; Kim et al., 2004
Nb		0–4	Goto et al., 1992; Collins, 2002
Co	1.1	0–10	Hashimoto et al., 1995; Collins, 2002; Sorano et al., 2004

TABLE 12.2

Elemental Composition of HiCr Rolls (in wt-%)

	Mill Rolls	Literature	References
C	1.9–3	0.8–3	Sano et al., 1992; Sorano et al., 2004; Belzunce et al., 2004
Ni		0.3–2	Hashimoto et al., 1995; Belzunce et al., 2004; Pellizzari et al., 2005
Cr	12.4–18.9	7–25	Sano et al., 1992; Hashimoto et al., 1995; Sorano et al., 2004
Mo	0.5–1.8	0.5–5.5	Hashimoto et al., 1995; Sorano et al., 2004; Belzunce et al., 2004
V	≤0.5	≤3	Sano et al., 1992; Hashimoto et al., 1995; Park et al., 1999
W		0	Goto et al., 1992; Sano et al., 1992; Sorano et al., 2004
Si	0.4–0.8	0.4–1	Lienard et al., 1995; Lecomte-Beckers et al., 1997; Park et al., 1999; Li et al., 2000
Mn	0.1–1.1	0.5–11	Park et al., 1999; Li et al., 2000; Kang et al., 2001; Milan et al., 2005

$M_{23}C_6$ (Goto, Matsuda, and Sakamoto, 1992; Molinari et al., 2000; Kang et al., 2001), whereas HiCr rolls contain mainly M_7C_3, with little MC, M_3C and M_2C (Li et al., 2000; Pellizzari et al., 2006; Joos et al., 2007). The content of individual carbides in HSS is shown in Table 12.4. HiCr rolls contain 22–34 vol-% of M_7C_3 carbides (Kang et al., 2001; Pellizzari, Molinari, and Straffelini, 2005; Pellizzari et al., 2006); ICDP rolls contain 26–35 vol-% of M_3C carbides, which are mainly F_3C (Belzunce, Ziadi, and Rodriguez, 2004; Pellizzari et al., 2005).

TABLE 12.3

Elemental Composition of IC Rolls (in wt-%)

	Mill Rolls	Literature	References
C	2.8–3.5	2.5–3.5	Sano et al., 1992; Lecomte-Beckers et al., 1997; Belzunce et al., 2004
Ni	4–4.7	3.5–6	Lecomte-Beckers et al., 1997; Belzunce et al., 2004
Cr	0.5–1.9	1–2.5	Sano et al., 1992; Lecomte-Beckers et al., 1997; Belzunce et al., 2004
Mo	0.3–0.5	0.2–1	Sano et al., 1992; Lecomte-Beckers et al., 1997; Belzunce et al., 2004
V+Nb		0–1.5	Belzunce et al., 2004
Mn	0.5–1	1	Lienard et al., 1995
Si	0.7–2	0.9–1.1	Lecomte-Beckers et al., 1997

12.3 Role of Key Elements

Xin and Perks (1999) provided an account of the role of key elements in HSS rolls:

1. *Vanadium and Niobium:* Both form very hard MC carbides. Their total content should be 4–7 wt-%, because segregation will occur at higher content during the solidification in the centrifugal process. According to Kang et al. (2001), VC is the first carbide formed, and remaining C then forms carbides with W, Mo and Cr.
2. *Tungsten:* Forms duplex carbides, and is usually present as Fe_4W_2C. These carbides have a fish-bone shape and cannot be broken easily. The content of 1.5–2.5 wt-% offers the best wear and cracking resistance, and ductility.
3. *Molybdenum:* Has a similar function as W, and its ideal content is 4–6 wt-%. Ikeda et al. (1992) observed that tensile strength decreases with it, without affecting hardness.
4. *Carbon:* Affects cracking resistance and reduces ductility if content exceeds 2 wt-%.
5. *Silicon:* With C helps fluidity during casting, and keeps good strength and ductility, with the recommended range of 0.3–1 wt-%.
6. *Manganese:* Minimizes the oxygen content of liquid HSS and combines with sulphur in MnS, preventing grain boundary embrittlement. The range should be 0.5–1 wt-%.

Carbide type		Morphology	Chemistry	Localization
MC		• Globular • Thick • Isolated or associated	• Mainly V • Secondary Mo, W, Cr	Centre of grains or grain boundaries (in association with M_7C_3)
M_2C		• Acicular (needles) or lamellar • Associated	• Mainly Mo, W • Secondary Cr, Fe, V	Interdendritic areas
M_6C		• Thin lamellae (fish bone) • Associated	• Mainly Mo, W • Secondary Cr, Fe, W, V	Areas of strong cooling (first 5 mm from surface)
M_7C_3		• Thick lamellae (fish bone) • Associated	• Mainly Fe, Cr • Secondary Mo, V, W	Interdendritic areas
$M_{23}C_6$		• Small globules • Isolated	• Mainly Cr, Fe • Secondary Mo, W, V	Homogeneously reparted in matrix

(a)

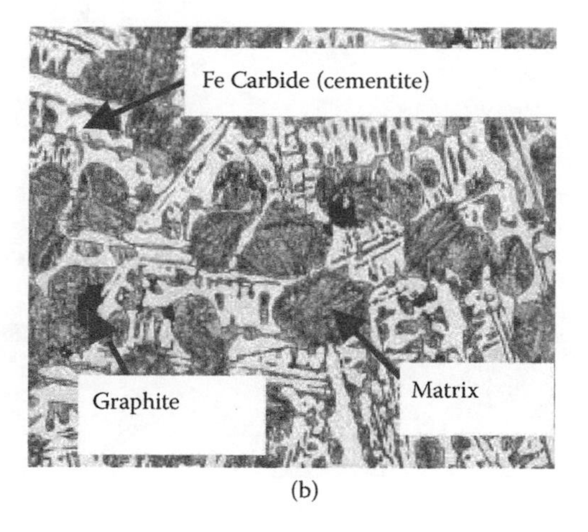

(b)

FIGURE 12.1
Summary of carbide morphology. The microstructure in the latter photo is included to show the shape of M_3C (white forms), the only significant carbide in ICDP rolls. [(a) Reprinted from V. Vitry, S. Nardone, J.-P. Breyer et al. *Materials and Design* 34:372–378, 2012. With permission from Elsevier. (b) From Sorano, H., N. Oda, and J.P. Zuccarolli, 2004, History of high-speed steel rolls in Japan, in *Proc. MS&T Conf.*, 26–29 September 2004, New Orleans, 379–390, Warrendale: The Minerals, Metals and Materials Society. Reprinted with permission of the MS&T sponsor societies.]

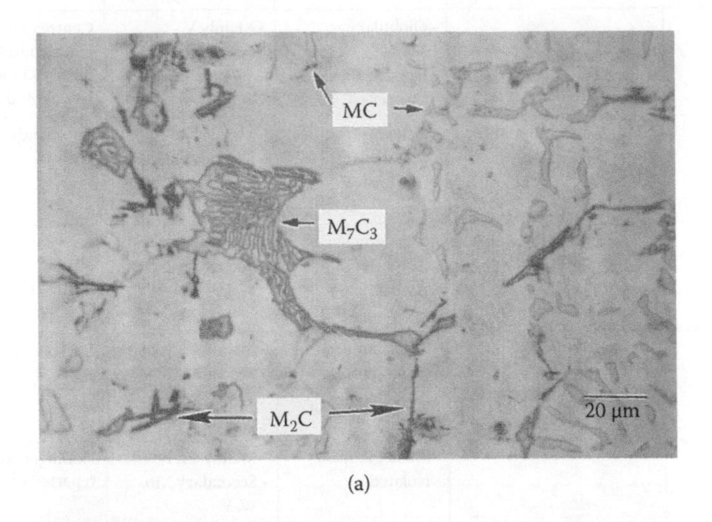

(a)

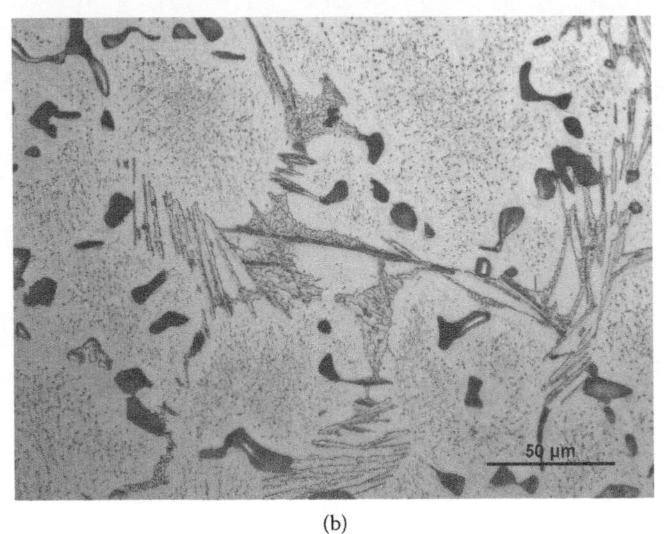

(b)

FIGURE 12.2
Appearance of carbides in HSS roll, [(a) Reprinted from S.J. Oh, S.-J. Kwon, H. Oh et al. *Met. Mat. Trans.* 31A: 793–798, 2000. With permission from Springer Verlag. (b) Courtesy of Dr Mario Boccalini, Jr]

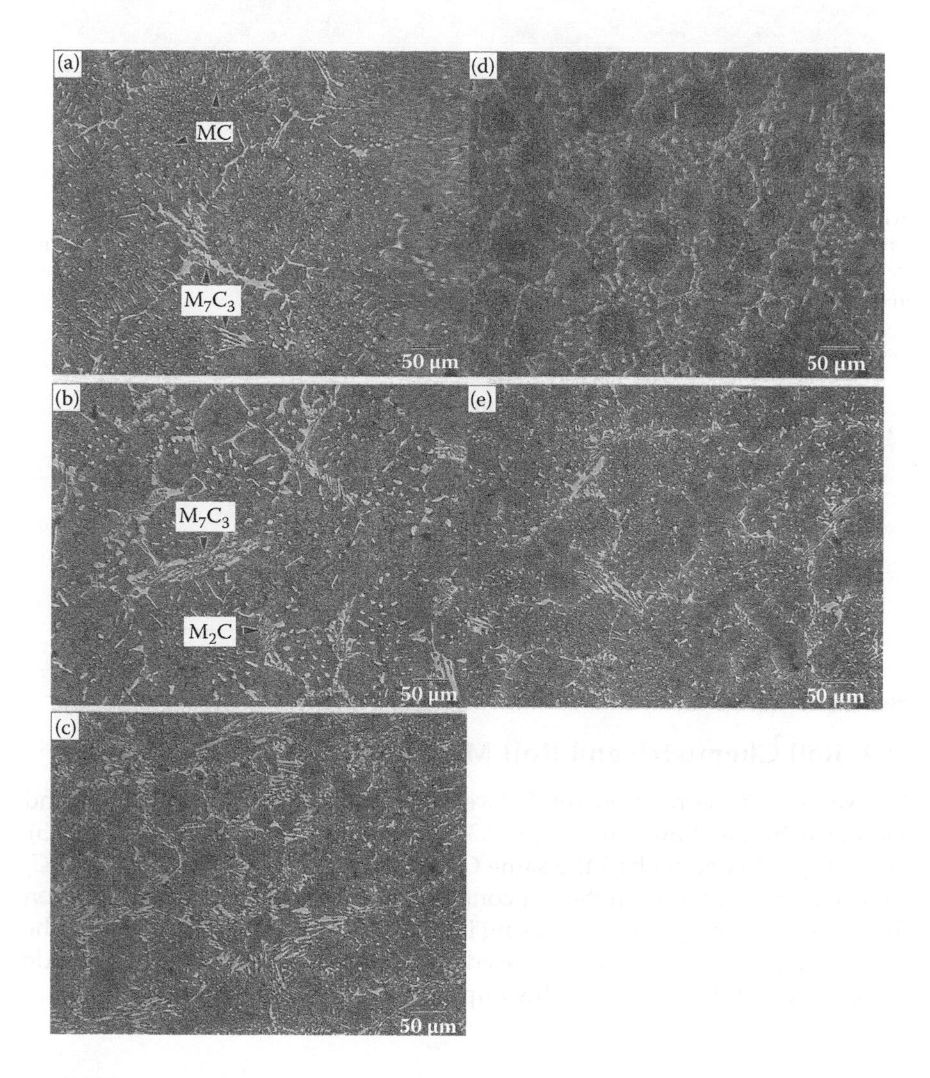

FIGURE 12.3
Five HSS samples. (Reprinted from Y.J. Kang, J.C. Oh, H.C. Lee et al. *Met. Mat. Trans. A* 32A:2515–2525, 2001. With permission from Springer Verlag.)

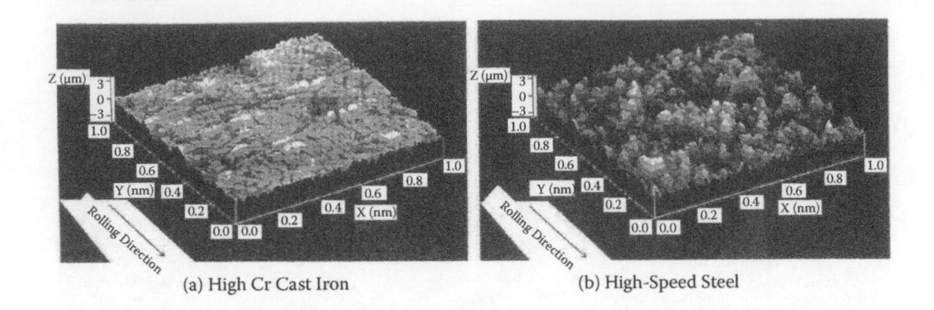

FIGURE 12.4
Topographic images of worn roll surface at HSS and HiCr rolls over an area of 1 × 1 mm². (Reprinted from K.Gotoh, H. Okada, T. Sasaki, et al. *Tetsu-to-Hagane* 84:861–867, 1998. With permission from Iron and Steel Institute of Japan.)

TABLE 12.4
Carbide Content of HSS Rolls in vol-%

	HSS	References
MC	**2–18**	Park et al., 1999; Kang et al., 2001; Pellizzari et al., 2005
M_2C	**2–8**	Xavier et al., 2005; Pellizzari et al., 2005
M_7C_3	**0.5–11**	Park et al., 1999; Kang et al., 2001; Pellizzari et al., 2005
Total	**7–30**	Park et al., 1999; Kang et al., 2001; Pellizzari et al., 2005
	13–28	Goto et al., 1992 (NB: Expressed as % of area)

12.4 Roll Chemistry and Roll Microstructure

An example of the relationship between the composition of roll shells and microstructure is shown in Figure 12.3 for different HSS samples (Table 12.5). Samples (a), (b) and (c) had the same C content, and the networking of M_7C_3 is more prominent with higher Cr content. Cr content has a strong impact on HiCr rolls; for example, by increasing its content from 14 wt-% to 17 wt-%, the volume fraction of matrix decreased from 62% to 58%, and M_7C_3 made the balance. An illustration of this impact is given later in Figure 15.7.

12.5 Microstructure and Its Impact on Friction

Early HSS rolls were reported to have a significantly higher friction than HiCr rolls because of the sharp protruding carbides, which is illustrated by

TABLE 12.5

Overview of Properties of Rolls in Figure 12.3

Sample	C[a]	Cr[a]	Volume Fraction of MC	Volume Fraction of M_7C_3[b]
a	2	3.9	15.1	4.1
b	2	6.3	13.5	6
c	2	8.6	11.8	8.8
d	1.7	6.2	12.9	4.4
e	2.4	6.3	13.2	8.9

Source: Kang et al., *Met. Mat. Trans. A* 32A:2515–2525, 2001. With permission.
[a] [wt-%].
[b] Includes a small amount of M_2C.

a photograph of worn surfaces on HiCr and HSS rolls (Figure 12.4). HiCr iron rolls developed a flat and smooth worn surface, whereas MC asperities formed on HSS rolls. This agrees with the observations by Park, Lee, and Lee (1999), where friction of HSS rolls increased with MC content, and decreased with M_7C_3 content. Kang et al. (2001) also noticed that friction increases with MC content.

Sorano et al. (2004) argued that early HSS rolls in Japan had a high COF caused by VC particles sticking out from the roll matrix. The solution was seen in replacing VC partially by eutectic MoC, which is network shaped, with hardness between that of VC and CrC. The impact of the change is shown in Figure 12.5, where low-friction HSS rolls had friction only slightly higher than HiCr rolls.

Thonus et al. (1997) also attributed this high friction to VC carbides sticking out of a much softer martensite matrix. The remedy was seen in lowering the hardness differential between carbides and matrix. Adding Cr, Mo and W to the matrix did not help, inasmuch as these elements formed carbides. Cobalt is a weak carbide former and stays in matrix, but it increased matrix hardness above 400°C little, and was costly.

On the other hand, Vergne et al. (2006) showed that HSS rolls have a higher COF than HiCr rolls in the range 20–650°C, but much lower at 950°C. Savage et al. (1996) observed that the COF of a HiCr sample on steel was between the COFs of two HSS samples of different, although unspecified, composition. Finally, in the commercial mill, HiCr rolls were found to have a higher COF than HSS rolls at the same stand (F2).

Schröder (2003) claimed that friction differs little between roll types, and contended that surface conditions, rather then roll material per se, determines friction. It is also claimed that oxide formation has more impact on friction than roll type. However, different roll materials produce different surface conditions. Results in Chapter 16 show that COF, and the oxide formation and retention depend strongly on the shell chemistry.

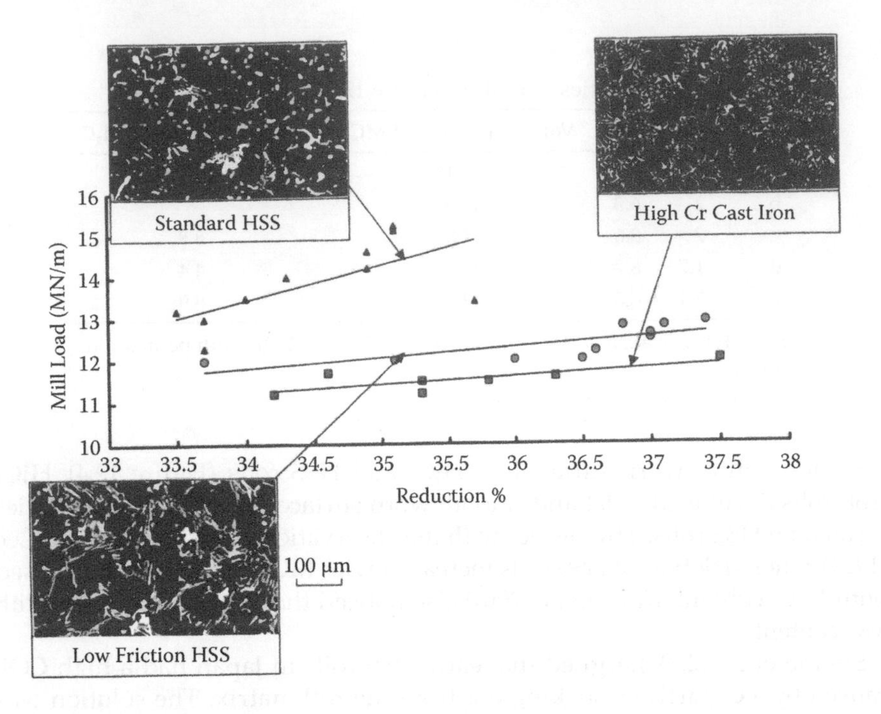

FIGURE 12.5
Effect of roll microstructure on the load, hence friction. (From Sorano, H., N. Oda, and J.P. Zuccarelli, 2004, History of high-speed steel rolls in Japan, in *Proc. MS&T Conf.*, 26–29 September 2004, New Orleans, 379–390, Warrendale: The Minerals, Metals and Materials Society. Reprinted with permission of the MS&T sponsor societies.)

12.6 Impact of Rare Earths and Silicon

Rare earths have a significant impact on roll properties. The addition of Ce (Yang et al., 2007), or a small amount of unspecified rare earths (Wang et al., 2007), refined the microstructure of HSS rolls (Figure 12.6) and increased their hardness by 20–30 HV. Duan, Jiang, and Fu (2007) observed a minuscule increase in the hardness of HSS rolls after adding Ce, but tensile strength increased by 16%. None of the authors, however, investigated the impact on friction. Luan et al. (2010) warned that this process is complicated, carbides may be coarser and the amount of unwanted inclusions may increase. It is also cautioned that the carbide networks may be eliminated in HSS rolls, which may increase friction. Another potential issue is that the portion of rod-shaped MC carbides increases with the additions of rare earths (Wang et al., 2011), which can increase friction.

Even a small amount of silicon on iron surfaces can have a significant, although not a clear-cut, impact on friction. Buckley and Brainard (1972)

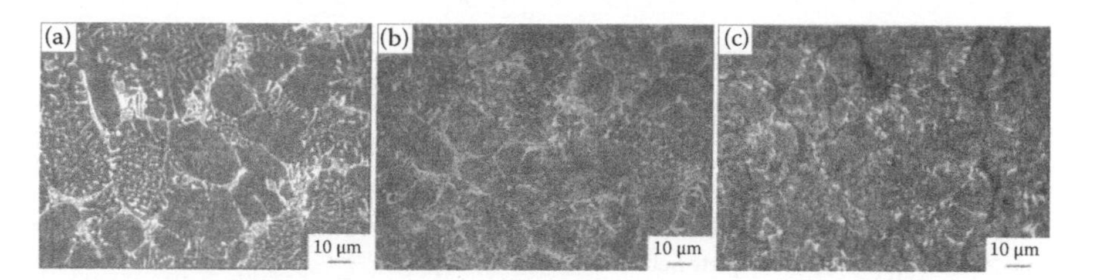

FIGURE 12.6
Effect of rare earths addition on the microstructure of HSS rolls; (a) no addition, (b) addition 0.04% and (c) addition 0.08%. (Reprinted from M. Wang, S. Mu, F. Sun, et al., *J. Rare Earths* 25:490–494, 2007. With permission from Elsevier.)

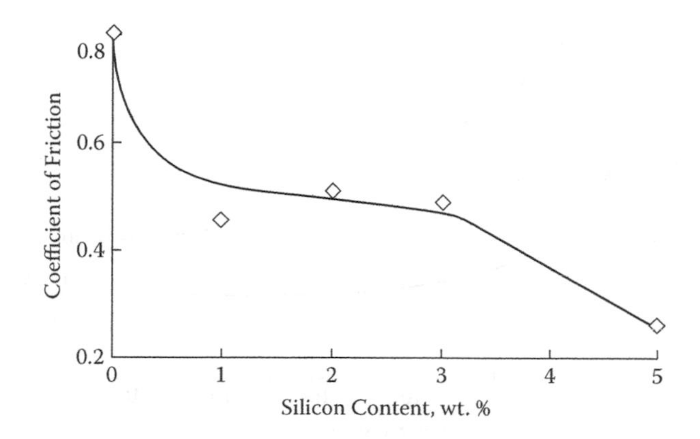

FIGURE 12.7
Effect of Si content on COF for Fe–Co sliding at 20°C in Ar. (Reprinted from D.H. Buckley, and W.A. Brainard, *Influence of Silicon on Friction and Wear of Iron-Cobalt Alloys*, 1972,NASA TN D-6769. With permission.)

reported that strain and heating enhance the segregation of iron-alloying elements at the surface, where their content may greatly exceed the bulk concentration. Silicon reduces the amount of metal-to-metal contact, thereby reducing friction (Figure 12.7). Buckley (1970) noticed that a small amount of oxygen reduces the friction much faster on Si-rich than on pure iron (Figure 12.8).

These observations were made in an oxygen-free environment. In the presence of oxygen, Genéve et al. (1999) found that silicon, which is more oxidisable than iron, segregates on the scale side of the iron/scale interface and hinders iron oxidation, resulting in thinner wüstite. Oike et al. (1992) investigated in laboratory conditions the effect of Si addition on the strip-based scale, and observed that friction increases with Si content of the strip.

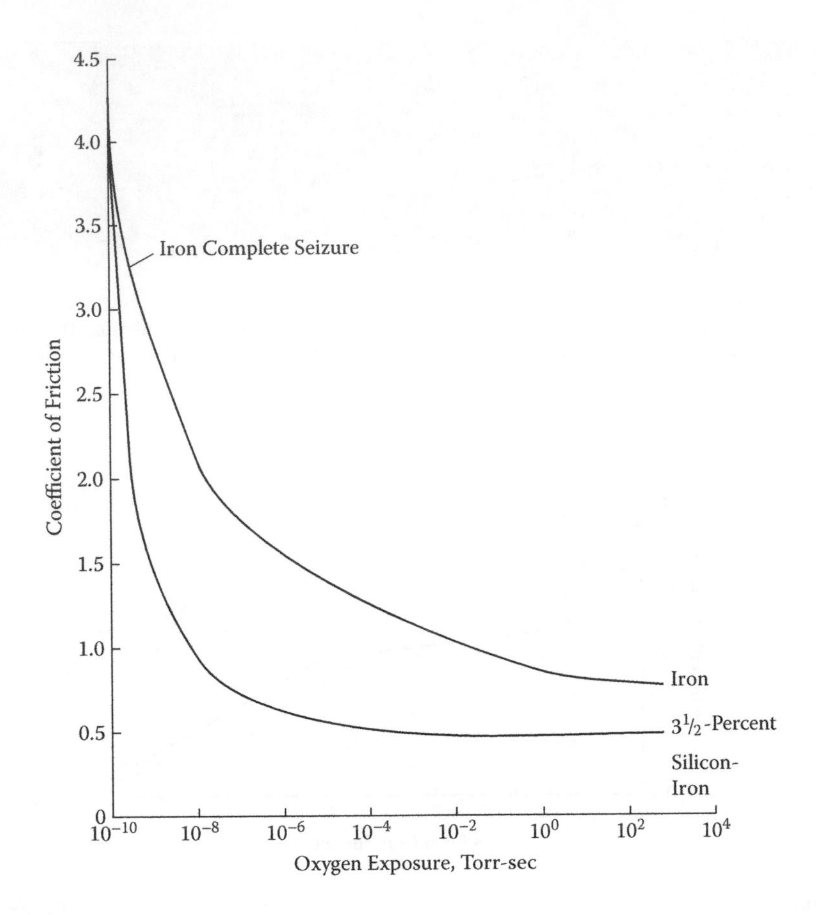

FIGURE 12.8
Effect of oxygen on COF of pure iron and 3.5 wt-% Si-Fe alloy in in vacuum. (Reprinted from D.H. Buckley, *Effect of 3½ Percent Silicon on Adhesion, Friction, and Wear of Iron in Various Media*, 1970, NASA TN D-5667.)

References

Belzunce, F.J., A. Ziadi, and C. Rodriguez. 2004. Structural integrity of hot strip mill rolling rolls. *Eng. Failure Anal.* 11:789–797.

Breyer, J.P., R.J. Skoczynski, and G. Walmag. 1997. High speed steel rolls in the hot strip mill. In *Proc. SARUC Conf., Vanderbijlpark.*

Buckley, D.H. 1970. *Effect of 3½ Percent Silicon on Adhesion, Friction, and Wear of Iron in Various Media*. NASA TN D-5667.

Buckley, D.H., and W.A. Brainard. 1972. *Influence of Silicon on Friction and Wear of Iron-Cobalt Alloys*. NASA TN D-6769.

Campbell, F.C. 2008. *Elements of Metallurgy and Engineering Alloys*. Materials Park: ASM International.

Collins, D. 2002. The metallurgy of high speed steel rolls. In *Rolls for Metalworking Industries*, G.E. Lee (Ed.), 83–91. Warrendale, PA: Association for Iron and Steel Technology.

Crime and the City Solution. 1990. The dolphins and the sharks. In *LP Paradise Discotheque*. Mute Records.

Duan, J., Z. Jiang, and H. Fu. 2007. Effect of RE-Mg complex modifier on structure and performance of high speed steel rolls. *J. Rare Earths* 25:259–264.

Garza-Montes-de-Oca, N.F., R. Colás, and W.M. Rainforth. 2011. On the damage of a work roll grade high speed steel by thermal cycling. *Eng. Failure Anal.* 18:1576–1583.

Genéve, D., M. Confente, D. Rouxel et al. 1999. Distribution around the oxide-substrate interface of alloying elements in low-carbon steel. *Oxid. Met.* 51:527–537.

Goto, K., Y. Matsuda, K. Sakamoto. 1992. Basic characteristics and microstructure of high-carbon high speed steel rolls for hot rolling mill. *ISIJ Int.* 32:1184–1189.

Gotoh, K., H. Okada, T. Sasaki et al. 1998. Effects of roll surface deteriorations on scale defect in hot rolling. *Tetsu-to-Hagane* 84:861–867.

Hashimoto, M., T. Kawakami, T. Oda et al. 1995. Development and application of high-speed tool steel rolls in hot strip rolling. *Nippon Steel Tech. Rep.* No. 66:82–90.

Higgins, R.A. 1983. *Engineering Metallurgy. Part 1: Applied Physical Metallurgy*, 5th ed. London: Edward Arnold.

Ikeda, M., T. Umeda, C.P. Tong et al. 1992. Effect of molybdenum addition on solidification structure, mechanical properties and wear resistivity of high chromium cast irons. *ISIJ Int.* 32:1157–1162.

Joos, O., C. Boher, C. Vergne et al. 2007. Assessment of oxide scales influence on wear damage of HSM work rolls. *Wear* 263:198–206.

Kang, Y.J., J.C. Oh, H.C. Lee et al. 2001. Effects of carbon and chromium additions on the wear resistance and surface roughness of cast high-speed steel rolls. *Met. Mat. Trans. A* 32A:2515–2525.

Kim, C.K., J.I. Park, J.H. Ryu et al. 2004. Correlation of microstructure and thermal-fatigue properties of centrifugally cast high-speed steel rolls. *Met. Mat. Trans.* 35A:481–492.

Lecomte-Beckers, J., J.T. Tchuindjang, R. Ernst et al. 2004. Structural investigations of HSS rolls for hot strip mill. In *Proc. 41st Rolling Seminar, Rolled and Coated Product*, 26–28 October 2004, Janville, Brazil.

Lecomte-Beckers, J., L. Terziev, and J.P. Breyer. 1997. Graphitisation in chromium cast iron. In *Proc. 39th MWSP Conf., XXXIV*, ISS, 423–431.

Lee, J.H., J.C. Oh, J.W. Park et al. 2001. Effects of tempering temperature on wear resistance and surface roughness of a high speed steel roll. *ISIJ Int.* 41:859–865.

Li, C.S., J.Z. Xu, X. He et al. 2000. Black oxide layer formation and banding in high chromium rolls. *Mat. Sci. Tech.* 16:501–505.

Lienard, P., C. Pacque, and J.-P. Breyer. 1995. Solutions modernes pour améliorer la qualité des cylindres de laminoirs moules. *Hommes et Fonderi* 259:23–28.

Luan, Y., N. Song, Y. Bai et al. 2010. Effect of solidification rate on the morphology and distribution of eutectic carbides in centrifugal casting high-speed steel rolls. *J. Mat. Proc. Tech.* 210:53–541.

Milan, J.C.G., M.A. Carvalho, R.R. Xavier et al. 2005. Effect of temperature, normal load and pre-oxidation on the sliding wear of multi-component ferrous alloys. *Wear* 259:412–423.

Molinari, A., G. Straffelini, A. Tomasi et al. 2000. Oxidation behaviour of ledeburitic steels for hot rolls. *Mat. Sci. Eng.* 280A:255–262.

Oh, S.J., S.-J. Kwon, H. Oh et al. 2000. Phase analysis of two steel work rolls using Mössbauer spectroscopy. *Met. Mat. Trans.* 31A: 793–798.

Oike, Y., J. Sato, and K. Minami et al. 1992. Influence of rolling conditions and chemical compositions of rolled material on strip surface flaw caused by surface deterioration of hot work rolls. *ISIJ Int.* 32:1211–1215.

Park, J.W., J.C. Lee, and S. Lee. 1999. Composition, microstructure, hardness and wear properties of high-speed steel rolls. *Met. Mat. Trans. A* 30A:399–409.

Pellizzari, M., A. Molinari, and G. Straffelini. 2005. Tribological behaviour of hot rolling rolls. *Wear* 259:1281–1289.

Pellizzari, M., D. Cescato, A. Molinari et al. 2006. Laboratory testing aimed at the development of materials for hot rolls. In *ATS Steel Rolling Conf.* Paris.

Sano, Y., T. Hattori, and M. Haga. 1992. Characteristics of high-carbon high speed steel rolls for hot strip mill. *ISIJ Int.* 32:1194–1201.

Savage. G., R. Boelen, A. Horti et al. 1996. Hot wear testing of roll alloys. In *Proc. 37th MWSP Conf.*, 333–337.

Schröder, K.H. 2003. *A basic understanding of the mechanics of rolling mill rolls.* Tenneck: Eisenwerke Sulzau-Werfen.

Sorano, H., N. Oda, and J.P. Zuccarelli. 2004. History of high speed steel rolls in Japan. In *Proc. MS&T Conf., 26–29 September 2004, New Orleans*, 379–390. Warrendale, PA: Minerals, Metals and Materials Society.

Thonus, P., J.C. Herman, J.P. et al. 1997. Off-line analysis of the HSS roll behaviour in the hot strip mill by the use of a rolling load prediction model. In *Proc. 38th MWSP Conf.*, 43–49. ISS.

Vergne, C., D. Batazzi, C. Gaspard et al. 2006. Contribution of laboratory tribological investigations on the performance appraisal of work rolls for hot strip mill. In *Proc. ATS Rolling Conf., Paris, June 2006.*

Virgil. 1904. *Georgics*. Boston: Riverside Press.

Vitry, V., S. Nardone, J.-P. Breyer et al. 2012. Microstructure of two centrifugal cast high speed steels for hot strip mills applications. *Mater. Des.* 34:372–378.

Wang, M., S. Mu, F. Sun et al. 2007. Influence of rare earth elements on microstructure and mechanical properties of cast high-speed steel rolls. *J. Rare Earths* 25:490–494.

Wang, M., Y. Li, Z. Wang et al. 2011. Effect of rare earth elements on the thermal cracking resistance of high speed steel rolls. *J. Rare Earths* 29:489–493.

Wittgenstein, L. 1922. *Tractatus logico-philosophicus*. London: Routledge and Kegan Paul Ltd.

Xavier, R.R., M.A. de Carvalho, E. Cannizza et al. 2005. Successful strategy for HSS rolls implementation. In *Roll Technology Mat. Sci. Tech. Conf.* Pittsburgh, R. Webber and P.C. Perry (Eds.), 55–64.

Xin, Z., and M.C. Perks. 1999. Production of HSS rolls for use in narrow hot strip mills and rod mills. In *Proc. Conf. 'Rolls 2000+.'* Birmingham.

Yang, J., D. Zou, X. Li et al. 2007. Effect of rare earth on microstructures and properties of high speed steel with high carbon content. *J. Iron Steel Res. Int.* 14:47–52.

13

Presence and Behaviour of Oxides in Roll Gap

In the absence of roll bite lubrication, oxides are the only major lubricants in hot rolling. Their formation on rolls and strip is addressed first, followed by a discussion of the composition and colour of oxides on rolls. Finally, the adherence of oxides to substrate, their impact on friction in roll gap and the modelling of their growth are analysed.

13.1 Mechanism of Oxide Formation

This part discusses a variety of topics: oxide formation on rolls and strip, transfer of oxide between rolls and strip, and the impact of moisture on oxide growth, which is important given the amount of cooling water applied to rolls and strip in a hot strip mill. It also attempts to answer whether HiCr rolls are more prone to oxidation than HSS rolls.

13.1.1 Rolls and Strip

It is generally agreed that oxidation of HSS rolls starts at the carbide/matrix interface, and some observed that the oxidation then progresses on the matrix, and then on carbides (Molinari et al., 2000; Pellizzari, Molinari, and Straffelini, 2005; see also Table 13.1). Zhu et al. (2010) observed that it first progresses over carbides, and then on the matrix.

Regarding carbide oxidation, MC has a much higher oxidation rate than other carbides, particularly M_2C and M_6C (Walmag, Skoczynski, and Breyer, 2001; Pellizzari et al., 2005; Joos et al., 2007; see also Fig. 13.1), which can be explained by low Cr content. On the other hand, the oxidation rate of carbide with high Cr content, M_7C_3, is very low (Pellizzari, Cescato, and Molinari, 2006; Joos et al., 2007). This is consistent with Xavier et al. (2005), who found that the oxidation rate for HSS without M_7C_3 was much higher (by ~40%) than for HSS containing it.

The comparative oxidation rate of matrix and carbides is a complex issue. Kim, Lim, and Lee (2003) noted that carbides MC and M_2C oxidise faster than matrix. However, Boccalini and Sinatora (2002) claimed the opposite, which agrees with the findings of Pellizzari et al. (2005). Finally, Walmag

TABLE 13.1
HSS and HiCr Sample Oxidation Scenario

HSS		HiCr	
Temp. (°C)	**Onset of Oxidation**	**Temp. (°C)**	**Onset of Oxidation**
350	Matrix interface with M_2C and M_7C_3	350	Matrix and M_2C/matrix interface
400	Matrix	400	M_7C_3/matrix interface
400–650	MC/matrix interface		
650	MC carbides		

Source: Extracted from Joos et al., *Wear* 263:198–206, 2007. With permission.

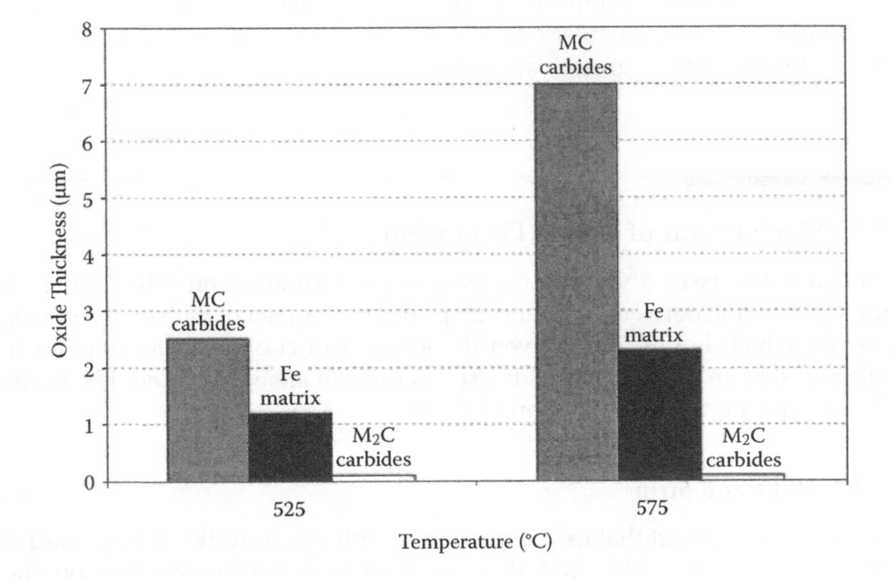

FIGURE 13.1
The influence of temperature on the oxidation of HSS. (Reprinted from G. Walmag, R.J. Skoczynski, and J.P. Breyer, *La Rev. Met.* 98:295–304, 2001. With permission from EDP Sciences.)

et al. (2001) argued that matrix oxidises much more slowly than MC and much faster than M_2C (Figure 13.1).

A possible cause of the discrepancies is the content of chromium and its portion tied up in carbides. Although Pellizzari et al. (2006) showed that the oxidation rate decreases with increasing Cr content (Figure 13.2), Cr alone does not determine the rate. At the same Cr content, HSS samples with high carbide content had a much higher oxidation rate than those with low carbide content. The so-called semi-HSS, with 2–4 times smaller C content than common HSS rolls, had a low rate, with no relationship to the Cr content. The low oxidation can be explained by the possibility that a large portion of Cr stayed in the matrix due to the lack of carbon for carbide formation.

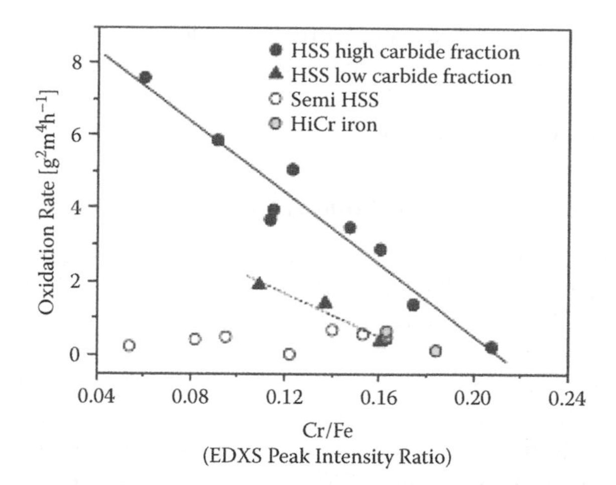

FIGURE 13.2
Oxidation rate as a function of Cr/Fe ratio. (Reprinted from M. Pellizzari, D. Cescato, A. Molinari et al., *ATS Steel Rolling Conference*, Paris, 2006. With permission from Fédération Française de l'Acier.)

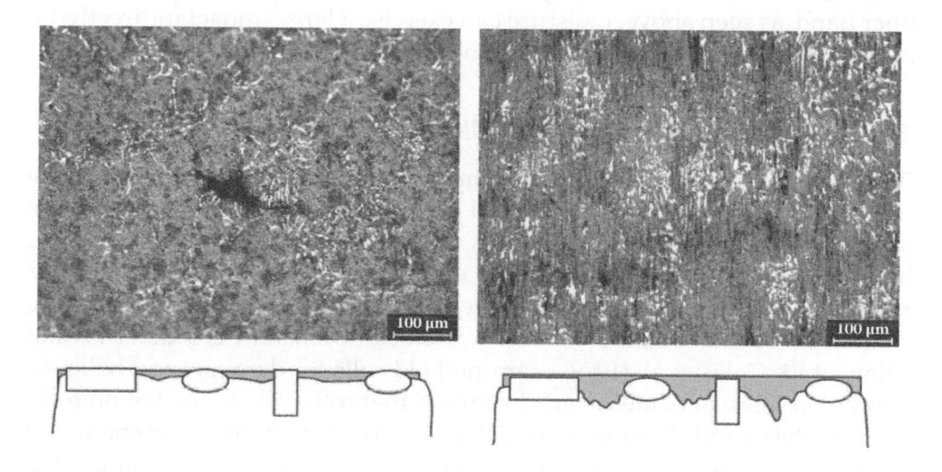

FIGURE 13.3
The oxide coverage of HSS (left) and HiCr rolls (right), with the surface appearance shown at the top, and the oxide–roll interface at the bottom. (Reprinted from M., Pellizzari, A. Molinari, and G. Straffelini, *Wear* 259:1281–1289, 2005. With permission from Elsevier.)

Werquin and Bocquet (1992) and Collins (2002) also contended that higher Cr content means less oxidation. However, Pellizzari et al. (2005) showed that the matrix can have a high oxidation rate, despite high bulk Cr content, if most Cr is bound in carbides (Figure 13.3). HiCr samples developed much thicker oxide than HSS samples, despite much higher Cr content (15–18 vol-% versus 4–6 vol-%) and the absence of MC carbides.

The impact of the strip temperature on oxide composition is shown in Figure 13.4, and those results agree with Basabe and Szpunar (2004), who recommended rolling strip at 850°C, because oxide consists mainly of soft wüstite, with very little hæmatite.

13.1.2 Impact of Moisture on Oxide Growth

Generally, the presence of moisture enhances the growth of oxide on both strip and rolls. Echsler, Ito, and Schütze (2003) noted that oxide on a low-C strip was thinner in a dry than in wet atmosphere. However, varying water-vapour content from 7%–19.5% made little impact on the thickness and morphology of scale. Kim, Lim, and Lee (2003) experimented with HSS rolls and found that both carbides and martensite matrix were oxidised in a dry atmosphere. Only the matrix was oxidised in wet conditions, but the oxide was thicker. Zhu and colleagues (2010) also experimented with HSS rolls, but observed that the presence of water vapour increases the oxidation rate of both carbides and matrix. Incidentally, Lancaster (1990) contended that the direct impact of water on friction is generally insignificant on engineering surfaces; the impact of humidity on wear is much more significant. On the other hand, as seen above, moisture can exercise a large impact indirectly, by influencing the rate of oxidation and the type and microstructure of oxides.

13.1.3 Where Is Oxide Found on Rolls Actually Formed?

There is a diversity of opinions as to the origin of oxide found on rolls. Most researchers agree that oxide is formed on rolls, but there is also evidence of oxide transfer between strip and rolls. Werquin and Bocquet (1992) argued that oxide is formed on the roll surface; different types of rolls at the same stand have very different colour after rolling. HSS roll is always black, whereas at a 7-stand mill HiCr roll was black at F1 and F2, blue at F3 and yellow at F4. Colás et al. (1999) examined old rolls for damage, and believed that the presence of oxides in cracks proves that roll oxide is created on rolls. On the other hand, Vergne et al. (2001) observed on a rig an exchange of oxide between the roll and the strip material. The presence of hæmatite suggests that some oxide might have been transferred from strip, given the high temperature of its formation (Figure 13.4).

Yarita (1984) quoted evidence for both cases, and suggested that oxide is formed by the corrosion reaction on the strip, due to the high temperature and high-pressure steam (Figure 13.5). The initial product, $Fe(OH)_2$ is transformed in Fe_3O_4 'as it is rubbed onto the roll surface'. Yarita also argued that oxide is formed on rolls, judging by the presence of chromium in it.

Roll surface temperature determines the formation of roll oxide. Quinn (1983) considered three possible temperatures at which a surface oxidises, namely the 'hot-spot' temperature at the contact areas, general surface temperature and an intermediate temperature. The temperature of hot spots depends

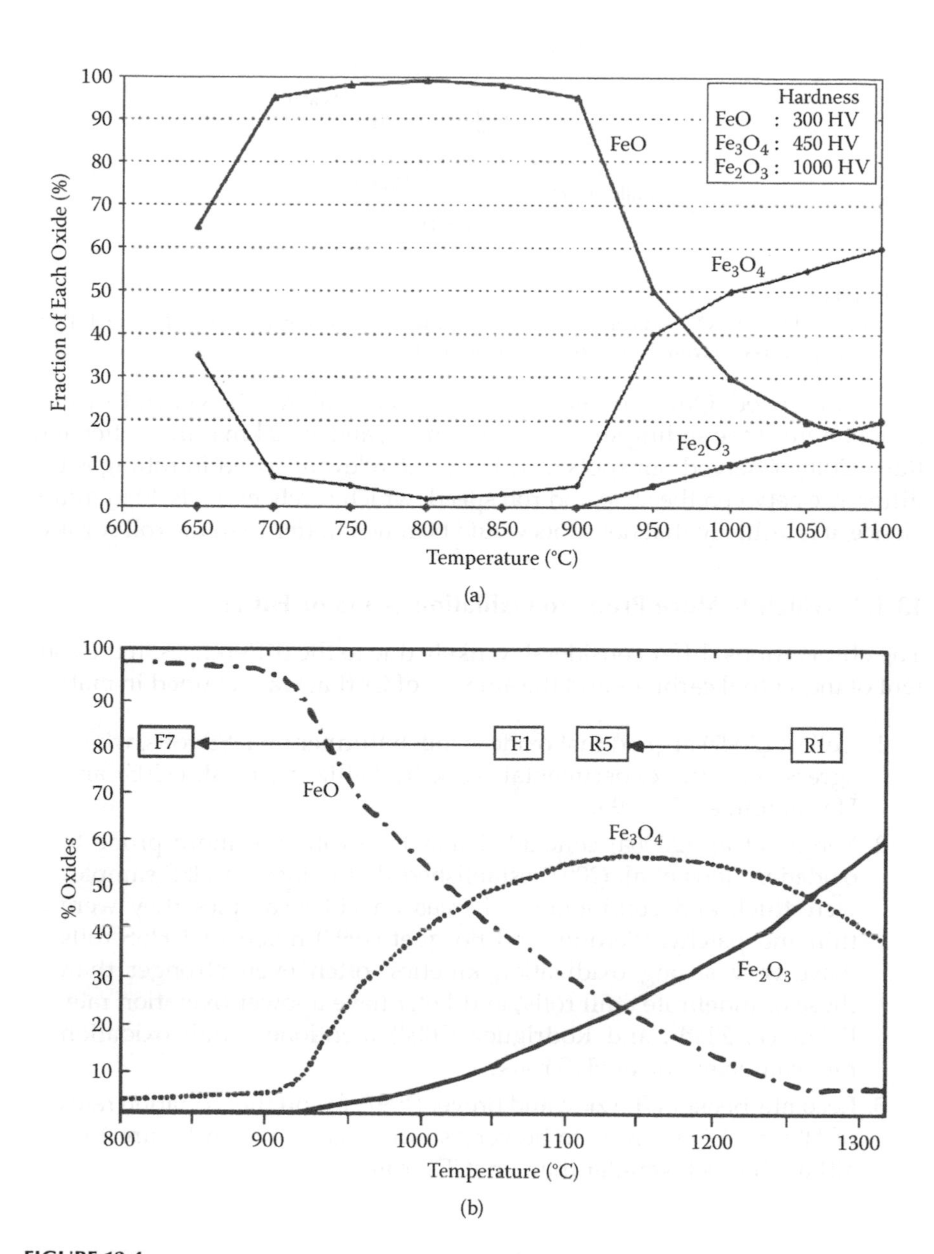

FIGURE 13.4
The effect of strip surface temperature on oxide composition. [(a) Reprinted from G. Walmag, R.J. Skoczynski, and J.P. Breyer, *La Rev. Met.* 98:295–304, 2001. With permission from EDP Sciences; (b) Reprinted from F.J. Belzunce, A. Ziadi, and C. Rodriguez, *Eng. Failure Anal.* 11:789–797, 2004. With permission from Elsevier.]

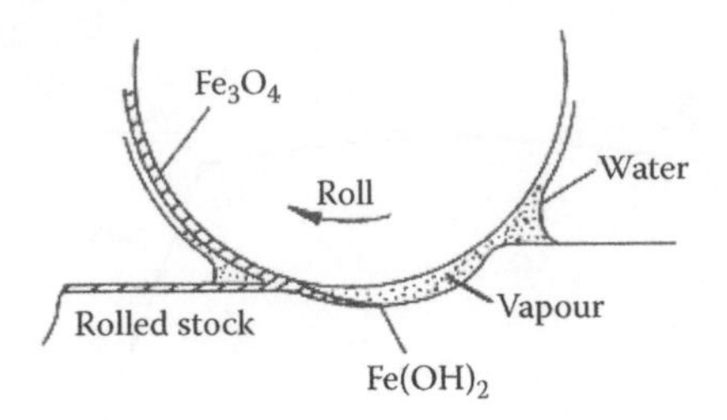

FIGURE 13.5
Formation of black Fe_3O_4 oxide on HiCr rolls. (Reprinted from I. Yarita, *Trans. ISIJ* 24:1014–1035, 1984. With permission from Iron and Steel Institute of Japan.)

on sliding speed. Quinn observed it to be ~200°C above the general surface temperature at the sliding speed of 375 m min^{-1}, and at ~2 km min^{-1} Sutter and Ranc (2010) observed hot spots of 1100°C. The relevant speed in rolling is the difference between the strip and roll speed, which rarely exceeds 40 m min^{-1}. Hence, it is unlikely that hot spots would be much hotter than the roll surface.

13.1.4 Which Is More Prone to Oxidation—HSS or HiCr?

The observations differ considerably, likely due to the differences in the content of individual carbides and the amount of Cr that was retained in matrix:

1. Collins (2002) argued that oxide is much thinner on HSS rolls, which agrees with the experimental work by Pellizzari et al. (2005) and Hashimoto et al. (1995).
2. Vergne et al. (2006a) concluded that HSS rolls are more prone to oxidation. Joos et al. (2007) established that oxides on HSS samples were thick and continuous, whereas on HiCr samples they were thin and patchy. Werquin and Bocquet (1992) noted that HSS rolls 'have very strong oxidization kinetics, often even stronger than those of indefinite chill rolls', and HiCr have a lower oxidation rate. Belzunce, Ziadi, and Rodriguez (2004) mentioned high oxidation rate 'characteristic of HSS rolls'.
3. Lecomte-Beckers, Terziev, and Breyer (1997) found the oxidation rates of HSS and HiCr rolls to be very similar, as shown in Figure 13.6, although much smaller than on ICDP rolls.

13.2 Properties of Oxide on Rolls

13.2.1 Composition

There is a divergence in the literature on the composition of oxide on rolls:

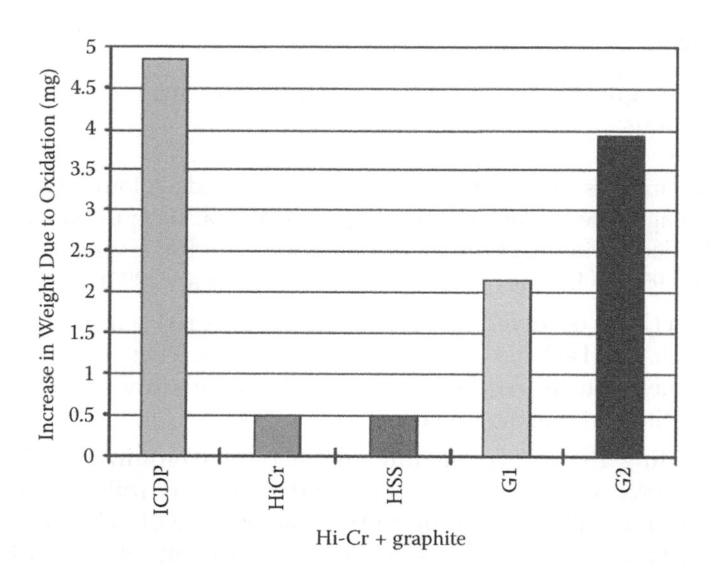

FIGURE 13.6
Oxidation behaviour of some roll types. (From Lecomte-Beckers, J., L. Terziev, and J.P. Breyer, 1997, Graphitisation in chromium cast iron, In *Proc. 39th MWSP Conf.*, ISS, 423–431. Association for Iron and Steel Technology. Reprinted with permission of the MS&T sponsor societies.)

1. Hashimoto et al. (1995) observed that a 5–15-μm-thick layer of hæmatite and magnetite forms on HSS rolls, 10–30 μm of hæmatite on HiCr rolls and a very thin layer of hæmatite and magnetite on high-Ni rolls.
2. Kim et al. (2003) also observed magnetite with some hæmatite on HSS rolls.
3. According to Molinari et al. (2000) and Pellizzari et al. (2006), M_3O_4 spinel (M = Fe,V, Cr) forms at the HSS roll surface, covered with hæmatite.
4. Beynon (1998) claimed that at temperatures below 570°C, magnetite forms next to metal, and hæmatite above it. Above 570°C, wüstite forms next to metal.
5. Walmag et al. (2001) suggested that at the usual roll surface temperatures (<500°C) most of the oxide would be magnetite, with little wüstite and no hæmatite (Figure 13.4). This is consistent with observations reported for HSS rolls by Caithness, Cox, and Emery (1999; magnetite with some wüstite), and Boccalini and Sinatora (2002; magnetite).
6. According to Morales, Sandoval, and Murillo (1999), iron and chromium oxides form on HiCr rolls.

Different surface temperatures could cause the large differences. Buckley (1983) gave an example of 440C bearing steel (1 wt-% C, 16–18 wt-% Cr, 0.75 wt-% Mo). At 600°C, iron oxide was dominant, but at 700°C chromium oxide dominated.

13.2.2 Colour

Generally, magnetite and wüstite are black, and hæmatite is red, with some variations:

1. Some authors give clear-cut definitions of oxide colours. For example, magnetite is black (Lancaster, 1963; Yarita, 1984; Li, Xu, and He, 2000; Eadie, Kalousek, and Chiddick, 2002), and so is wüstite (Bisson et al., 1956), whereas hæmatite is red (Eadie et al., 2002).
2. Quinn (1983) observed that α-Fe_2O_3 is black or red; a mixture of 69% α-Fe_2O_3, 24% FeO and 2.5% Fe was dark brown. Jones (1985) reported that hæmatite is reddish-orange and the mixture of magnetite, hæmatite and wüstite is black.
3. Regarding the roll colour after extraction, Werquin and Bocquet (1992) gave a detailed description of the colour of rolls after extraction in Table 13.2. According to Walmag et al. (2001), HSS rolls at F1 and F2 are covered in black oxide, and HiCr rolls at F3 and F4 are grey with uniformly distributed spots of black oxide.
4. According to Martiny (1998), the colour of oxide film is linked to its thickness. Yellow, blue and dark blue indicate the scale thickness of ~0.4 μm, 0.7 μm and 1 μm, respectively.

The appearance of HiCr rolls in the commercial mill was interesting, because a thin patchy layer of reddish-orange oxide, possibly hæmatite, covered the high-friction rolls. The lower-friction rolls were covered with uniform black oxide, likely magnetite and/or wüstite. The possible cause is discussed in Section 15.1.

TABLE 13.2

Colour of Different Roll Types at Stands F1–F7

	Stands						
Roll Type	**F1**	**F2**	**F3**	**F4**	**F5**	**F6**	**F7**
ICDP	Black	Black	Black	Black	Blackish blue	Blue/yellow	Light yellow
Adamite	Black	Black	Black				
HiCr iron	Black	Black	Blue	Yellow			White
HiCr steel	Blackish blue	Light blue	Yellow	White			
HSS	Black	Black	Black	Black	Black	Blue/yellow	Blue/yellow

Source: Reprinted from J.-C. Werquin and J. Bocquet, *Proc. 34th MWSP Conf.*, 1992, ISS, 135–151. With permission from the Association for Iron and Steel Technology.

13.2.3 Constancy

Oxide film breaks up for several reasons: difference in the thermal expansion of oxide and substrate, shear stress and pressure. The observations of the impact of the break-up on friction are somewhat divergent:

1. There is anecdotal evidence that broken oxide reduces friction, its particles acting like ball bearings. Schey (1983) suggested that friable oxides reduce friction. Munther and Lenard (1997) also proposed that broken oxide provides lubricity. Hinsley, Male, and Rowe (1968) noted that ZnO has high friction if it is not pulverised.
2. On the other hand, Bailey and Godfrey (1954) observed that this break-up and generation of loose particles increase friction, which is consistent with Stott (2002) and Das, Palmiere, and Howard (2004). Vergne et al. (2006b) also stated that compacted oxide has lower friction than the one consisting of loose particles. If oxide is loose, the wear mechanism is abrasion, and friction is high; if compacted, wear is by adhesion and friction is low. It is possible that loose particles are easily oxidised into abrasive hæmatite, increasing COF.
3. Loss of oxide from the roll surface due to banding leads to a serious increase in friction (see Chapter 14).

13.3 Impact of Oxides on Friction in Roll Gap

13.3.1 Thickness of Oxide

Oxide thickness is definitely related to friction, although observations differ:

1. Azushima and Nakata (2010) investigated COF in hot rolling of 9-mm-thick Si-Mn steel strip, with lighter and heavier lubrication. Figure 13.7 shows that (a) COF increases with reduction, although impact is smaller with thicker scale; (b) COF decreases with increased lubrication; and (c) COF initially decreases with oxide thickness, then stays steady or increases slowly (*cf.* Figure 4.4), except at light lubrication and heavy reduction.
2. Koseki, Yoshida, and Inoue (1994) and Munther and Lenard (1997) contended that thicker oxide reduces COF, but Kang et al. (2001) and Garza-Montes-de-Oca and Rainforth (2009) asserted the opposite.
3. In tests with lubrication COF decreased with oxide thickness, but oxide thickness had no effect in the unlubricated case (Sun, 2005).

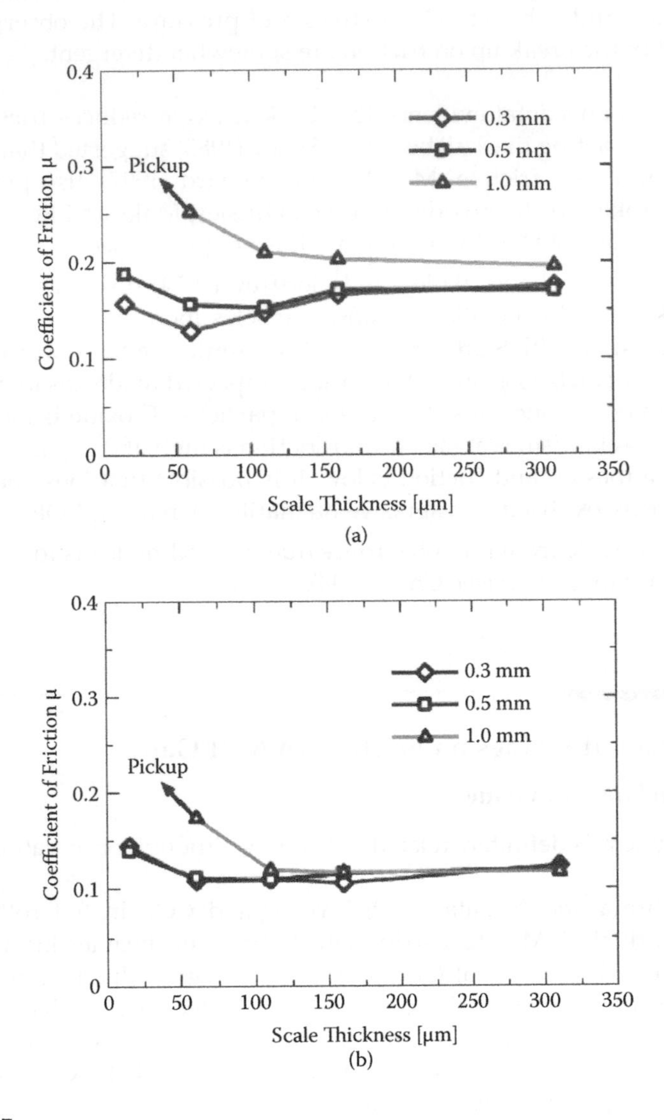

FIGURE 13.7
The effect of scale thickness, degree of lubrication and strip thickness reduction on friction in hot rolling of steel sheet, with (a) light, and (b) heavier lubrication. Legend indicates the reduction. (Reprinted from A. Azushima and Y. Nakata, *ISIJ Int.* 50:1447–1452, 2010. With permission from Iron and Steel Institute of Japan.)

4. Some tests with HSS and HiCr did not show any significant impact of pre-oxidation on COF (Milan et al., 2005). However, pre-oxidation reduced COF of HSS material in the temperature range of 20°C–950°C (Vergne et al., 2006a). Interestingly, it reduced the COF of HiCr samples at 650°C, but slightly increased COF at 950°C.

These discrepancies could be due to a number of factors, some of which are given in Section 4.4. As can be seen in Figure 4.4, COF tends to decrease with thickness up to a certain point, and then the trend reverses. The results cited above could have been influenced by the oxide thickness present in individual cases.

Jarl (1993) postulated the following impact of scale thickness on friction:

1. If strip scale is thin, it will quickly cool down towards roll temperature in the roll gap.
2. Cold scale may be harder than the strip, and will not lubricate.
3. Thicker scale will not cool that fast in contact with rolls. It may stay hot and softer than strip, decreasing friction.

13.3.2 Relative Hardness of Oxide, Strip and Roll

As discussed in Section 4.4, a layer of soft material between two harder surfaces reduces COF. The relative hardness of strip, oxide and roll play a critical role, and is shown in Figure 13.8 and Appendix D. Carbides are much harder

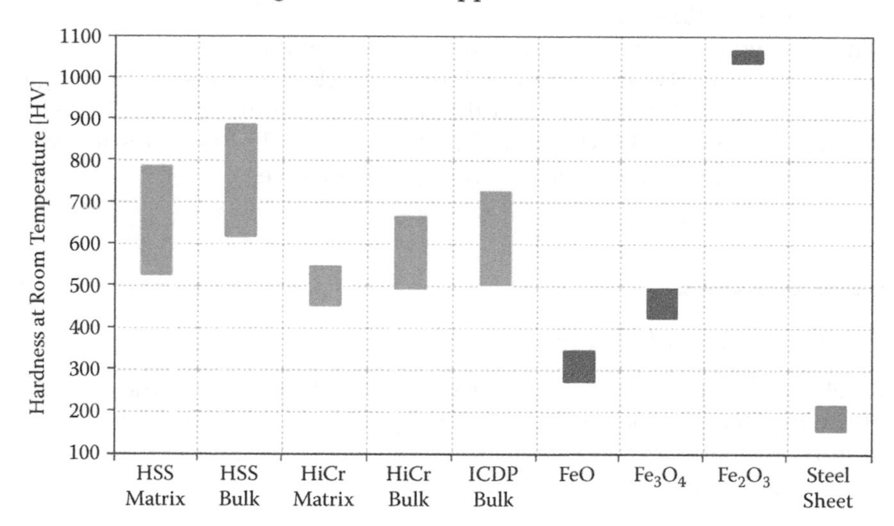

FIGURE. 13.8
Hardness (at 20°C) of rolls. [Compiled from the data in Park et al., 1999; Xin and Perks, 1999; Kang et al., 2001; Pellizzari et al., 2005; Ref. Marichal Ketin, ICDP), iron oxides (Lundberg and Gustaffson, 1994) and steel (Ref. Hot rolled hardness).]

TABLE 13.3

Pressure at Which Iron Oxides May Melt

Oxide	Latent Heat of Fusion (cal g^{-1})	Density (g cm^{-3})	Melting Temperature (K)	Pressure (MPa)
Wüstite	80	5.7	1650	206
Magnetite	142.5	5.17	1870	372
Hæmatite	112	5.1	1730	270

than other species in roll gap, and any iron oxide is a lubricant on them. Magnetite and, particularly, wüstite are lubricants on both the roll matrix and the bulk surface. Hæmatite is so hard that it can be classified as an abrasive in roll gap.

Hardness of these species would decrease at elevated temperatures experienced during hot rolling. At roll surface temperature of 500°C, hardness of oxides would be 140 HV, 450 HV and 660 HV for wüstite, magnetite and hæmatite, respectively. Based on the data in Appendix D, wüstite and, likely, magnetite would remain softer than rolls, and act as lubricants.

13.3.3 Liquefaction of Oxide

Windhager (2010) speculated that iron oxide liquefies under pressure, acting as a liquid lubricant. Olsson et al. (1998) claimed the opposite, that is, that at high velocities and pressures liquid lubricants solidify. It is advisable to check if the oxides can actually liquefy, and the required pressure was calculated using Equation (9.2).

The data were selected from Appendix D, and for the roll surface temperature of 300°C, results are shown in Table 13.3. Median pressure in roll gap was calculated with the data from the commercial mill, and was 208 MPa, 314 MPa, 340 MPa, 445 MPa and 545 MPa at stands F1–F5, respectively. Therefore, melting of iron oxides in roll gap is possible.

13.4 Modelling of Oxide Growth

Modelling of oxide growth on the strip is straightforward, and the approach of Browne, Dryden, and Assefpour (1995) performed well in the models for the hot strip rolling (Panjkovic, 2007) and reheat furnace (Panjkovic and Gloss, 2012). It is based on parabolic oxidation:

$$S_0 = \sqrt{At\exp(B/T)} \tag{13.1}$$

where S_0 is the scale thickness (m), t time (s) and the constants are $A = 2.95 \times 10^{-6}$ m^2 s^{-1} and $B = -1.5\ 10^{-4}$ K^{-1}. Oxidation occurs on a surface with changing temperature, and the modelling is based on the concept of equivalent time. It is the time during which the same amount of scale would be created at a standard temperature, in this case of 1000°C. For a time increment Δt, at a temperature T (K), the equivalent time is:

$$\Delta t_{eq} = \Delta t \exp\left(\frac{B}{T} - \frac{B}{1273}\right) \tag{13.2}$$

At each time step, the equivalent time is added to the total equivalent time and the scale thickness is calculated. Modelling of the oxide growth on rolls is more complicated because of the variety of species on the roll surface and the differences in their oxidation. As mentioned in Section 13.1, oxidation rates differ among various types of carbide. González et al. (2001) based their model on the assumption that the oxide thickness grows exponentially with time. They conducted experiments with HiCr samples to determine the oxidation rate and the exponent that would ensure the best fit between the measured and the calculated data. From the experiments conducted at 400°C, 500°C and 600°C, the formula for scale thickness can be interpolated as:

$$S_0 = 5.91 \times 10^{-5} \exp(-385/T)\, t^{0.21} \tag{13.3}$$

On the other hand, Li, Xu, and He (2000) proposed another model for the oxide growth on HiCr rolls. It was assumed that oxide grows between the exit from roll gap and the water cooling section, and that the travel time of the roll through that region is t^* (s). Oxide thickness is given by:

$$S_0 = 1.2 \times 10^{-4} t_R \left\{ 24.5n \int_0^{t^*} \exp\left[-\frac{10{,}190}{0.83T_S - 105.48\exp(s/0.00195)} \right] ds \right\}^{1/2} \tag{13.4}$$

where T_S (K) is the strip temperature, t_R (h) the rolling time, and n the number of roll revolutions per minute. Due to their relative simplicity, the first two models above were combined and included in the friction model described in Section 17.3.

References

Azushima, A., and Y. Nakata. 2010. Effect of scale on coefficient of friction in hot sheet rolling of steel. *ISIJ Int*. 50:1447–1452.

Bailey, J.M., and D. Godfrey. 1954. *Coefficient of Friction and Damage to Contact Area During the Early Stages of Fretting. III–Steel, Iron, Iron Oxide, and Glass Combinations.* Washington: NACA TN 3144.

Basabe, V.V., and J.A. Szpunar. 2004. Growth rate and phase composition of oxide scales during hot rolling of low carbon steel. *ISIJ Int.* 44:1554–1559.

Belzunce, F.J., A. Ziadi, and C. Rodriguez. 2004. Structural integrity of hot strip mill rolling rolls. *Eng. Failure Anal.* 11:789–797.

Beynon, J.H. 1998. Tribology of hot metal forming. *Tribol. Int.* 31:73–77.

Bisson, E.E., R.L. Johnson, M.A. Swikert et al. 1956. *Friction, Wear and Surface Damage of Metals as Affected by Solid Surface Films*. Washington: NACA Report 1254.

Boccalini, M. Jr, and A. Sinatora. 2002. Microstructure and wear resistance of high speed steels for rolling mill rolls. In *Proc. 6th Int. Tooling Conf., Karlstad, Vol. 1*, 425–438.

Browne, K.M., J. Dryden, and M. Assefpour. 1995. Modelling scaling and descaling in hot strip mills. In *Recent Advances in Heat Transfer and Micro-Structure Modelling for Metal Processing, MD-Vol. 67*, R.-M. Guo and J.J.M. Too (Eds.). ASME.

Buckley, D.H. 1983. Importance and definition of materials in tribology. In *Tribology in the 80's, Vol. 1*, 19–44. NASA CP-2300.

Caithness, I., S. Cox, and S. Emery. 1999. Surface behaviour of HSS in hot strip mills. In *Proc. Rolls 2000+ Advances in Mill Rolls Technology Conf.*, 111–120. Birmingham: Institute of Materials.

Colás, R., J. Ramirez, I. Sandoval et al. 1999. Damage in hot rolling work rolls. *Wear* 230:56–60.

Collins, D. 2002. The metallurgy of high speed steel rolls. In *Rolls for Metalworking Industries*, G.E. Lee (Ed.), 83–91. Warrendale, PA: Association for Iron and Steel Technology.

Das, S., E.J. Palmiere, and I.C. Howard. 2004. The cut-groove technique to infer interfacial effects during hot rolling. *Met. Mat. Trans.* 35A:1087–1095.

Eadie, D.T., J. Kalousek, and K.C. Chiddick. 2002. The role of high positive friction (HPF) modifier in the control of short pitch corrugations and related phenomena. *Wear* 253:185–192.

Echsler, H., S. Ito, and M. Schütze. 2003. Mechanical properties of oxide scales on mild steel at 800 to 1000°C. *Oxid. Metals* 60:241–269.

Garza-Montes-de-Oca, N.F., and W.M. Rainforth. 2009. Wear mechanisms experienced by a work roll grade high speed steel under different environmental conditions. *Wear* 267:441–448.

González, V., P. Rodriguez, Z. Haduck et al. 2001. Modelling oxidation of hot rolling work rolls. *Ironmaking and Steelmaking* 28:470–473.

Hashimoto, M., T. Kawakami, T. Oda et al. 1995. Development and application of high-speed tool steel rolls in hot strip rolling. *Nippon Steel Tech. Rep.* No. 66:82–90.

Hinsley, C.F., A.T. Male, and G.W. Rowe. 1968. Frictional properties of metal oxides at high temperatures. *Wear* 11:233–238.

Hot rolled hardness. http://steelproducts.bluescopesteel.com.au/home/steel-products/hot-rolled-coil (see XK15B28 XLERCOIL®; accessed April 21, 2013).

Jarl, M. 1993. An estimation of the mechanical properties of the scale at hot rolling of steel. In *1st Int. Conf. Model. Metal Rolling Proc.*, 21–23 September 1993, London, 614–628.

Jones, W.R. Jr. 1985. Boundary lubrication–Revisited. In *Tribology: The Story of Lubrication and Wear*, 23–53. NASA TM-101430.

Joos, O., C. Boher, C. Vergne et al. 2007. Assessment of oxide scales influence on wear damage of HSM work rolls. *Wear* 263:198–206.

Kang, Y.J., J.C. Oh, H.C. Lee et al. 2001. Effects of carbon and chromium additions on the wear resistance and surface roughness of cast high-speed steel rolls. *Met. Mat. Trans. A* 32A:2515–2525.

Kim, H., J.-W. Lim, and J.-J. Lee. 2003. Oxidation behaviour of high-speed steels in dry and wet atmosphere. *ISIJ Int.* 43:1983–1988.

Koseki, S., H. Yoshida, and K. Inoue. 1994. Improvement of accuracy of mathematical models for gauge set-up in hot strip finishing mills. *Tetsu-to-Hagane* 80:31–36.

Lancaster, J.K. 1963. The formation of surface films at the transition between mild and severe metallic wear. *Proc. R. Soc. A* 273:466–483.

Lancaster, J.K. 1990. A review of the influence of environmental humidity and water on friction, lubrication and wear. *Tribol. Int.* 23:371–389.

Lecomte-Beckers, J., L. Terziev, and J.P. Breyer. 1997. Graphitisation in chromium cast iron. In *Proc. 39th MWSP Conf.*, ISS, 423–431.

Li, C.S., J.Z. Xu, X. He et al. 2000. Black oxide layer formation and banding in high chromium rolls. *Mat. Sci. Tech.* 16:501–505.

Lundberg, S.-E., and T. Gustaffson. 1994. The influence of rolling temperature on roll wear, investigated in a new high temperature test rig. *J. Mat. Proc. Tech.* 42:239–291.

Marichal Ketin ICDP. http://www.mkb.be/index2.html (accessed April 21, 2013).

Martiny, F. 1998. Importance of roll cooling in HSS work rolls of hot strip mills. http://pandrinath.tripod.com/Gyanmanch.htm (accessed April 21, 2013).

Milan, J.C.G., M.A. Carvalho, R.R. Xavier et al. 2005. Effect of temperature, normal load and pre-oxidation on the sliding wear of multi-component ferrous alloys. *Wear* 259:412–423.

Molinari, A., G. Straffelini, A. Tomasi et al. 2000. Oxidation behaviour of ledeburitic steels for hot rolls. *Mat. Sci. Eng.* 280A:255–262.

Morales, J., I. Sandoval, and G. Murillo. 1999. Influence of process parameters on friction coefficient of high-chromium rolls. *AISE Steel Tech.* 76 (11):46–48.

Munther, P.A., and J.G. Lenard. 1997. A study of friction during hot rolling of steels. *Scand. J. Met.* 26:231–240.

Olsson, H., Åström, K.J., C. Canudas-de-Wit et al. 1998. Friction models and friction compensation. *Eur. J. Control* 4:176–195.

Panjkovic, V. 2007. Model for prediction of strip temperature in hot strip steel mill. *App. Therm. Eng.* 27:2404–2414.

Panjkovic, V., and R. Gloss, R. 2012. Fast dynamic heat and mass balance model of walking beam reheat furnace with two-dimensional slab temperature profile. *Ironmaking and Steelmaking* 39:190–211.

Pellizzari, M., A. Molinari, and G. Straffelini. 2005. Tribological behaviour of hot rolling rolls. *Wear* 259:1281–1289.

Pellizzari, M., D. Cescato, A. Molinari et al. 2006. Laboratory testing aimed at the development of materials for hot rolls. In *ATS Steel Rolling Conf.* Paris.

Quinn, T.F.J. 1983. *NASA Interdisciplinary Collaboration in Tribology. A Review of Oxidational Wear*. NASA CR 3686.

Schey, J.A. 1983. *Tribology in Metalworking. Friction, Lubrication and Wear*. Ohio: American Society for Metals.

Stott, F.H., 2002. High-temperature sliding wear of metals. *Trib. Int.* 35:489–495.

Sun, W. 2005. *A Study on the Characteristics of Oxide Scale in Hot Rolling of Steel*. PhD diss., University of Wollongong.

Sutter, G., and N. Ranc. 2010. Flash temperature measurement during dry friction process at high sliding speed. *Wear* 268:1237–1242.

Vergne, C., C. Boher, C. Levaillant et al. 2001. Analysis of the friction and wear behaviour of hot work tool scale: Application to the hot rolling process. *Wear* 250:322–333.

Vergne, C., D. Batazzi, C. Gaspard et al. 2006a. Contribution of laboratory tribological investigations on the performance appraisal of work rolls for hot strip mill. In *Proc. ATS Rolling Conf.*, Paris, June 2006.

Vergne, C., C. Boher, R. Gras et al. 2006b. Influence of oxides on friction in hot rolling: Experimental investigations and tribological modelling. *Wear* 260:957–975.

Walmag, G., R.J. Skoczynski, and J.P. Breyer. 2001. Improvement of the work roll performance on the 2050 mm hot strip mill at ISCOR Wanderbijlpark. *La Rev. Met.* 98:295–304.

Werquin, J.-C., and J. Bocquet. 1992. The new generation of spun cast rolls in high speed steels for hot strip mills. In *Proc. 34th MWSP Conf.*, ISS, 135–151.

Windhager, M. 2010. Private communication.

Xavier, R.R., M.A. de Carvalho, E. Cannizza et al. 2005. Successful strategy for HSS rolls implementation. In *Roll Technol. Mat. Sci. Tech. Conf.* Pittsburgh, R. Webber and P.C. Perry (Eds.), 55–64.

Xin, Z., and M.C. Perks. 1999. Production of HSS rolls for use in narrow hot strip mills and rod mills. In *Proc. Conf. "Rolls 2000+"*. Birmingham.

Yarita, I. 1984. Problems of friction, lubrication, and materials for rolls in rolling technology. *Trans. ISIJ* 24:1014–1035.

Zhu, Q., H.T. Zhu, A.K. Tieu et al. 2010. In-situ investigation of oxidation behaviour in high-speed steel roll material under dry and humid atmospheres. *Corrosion Sci.* 52:2707–2715.

14

Impact of Roll Wear on Friction

Different types of roll wear, which are addressed first, influence friction differently, as shown with the data collected with an online system for roll surface monitoring. Given that the fire-cracking is a widespread form of roll wear, it is discussed in some detail.

14.1 Basic Types of Roll Wear

There are many types of wear, and of interest here are those generated by the interaction of solids with contiguous surfaces (as opposed to the interaction of a solid surface with a swarm of particles, such as dust). The key types of wear of interest are as follows (Stachowiak and Batchelor, 2005):

1. Abrasive, where hard particles of one surface gouge the material from another surface.
2. Adhesive, where surfaces adhere to each other via asperity contacts, and chunks of weaker material are transferred to the stronger one at the separation of asperities.
3. Corrosive, due to chemical reactions at the interface. Oxidation wear is a special case, where oxidation by air occurs. A high rate of corrosion means the loss of metal substrate to the chemically formed film. The low rate, on the other hand, results in a thin film that can be broken, leading to adhesion between bare surfaces and adhesive wear.
4. Fatigue, due to the cyclic contact between surfaces during which high stresses are generated, followed by the relaxation of these stresses. Fatigue generates and propagates cracks, eventually leading to the loss of surface material.

Each of these types is caused by several different mechanisms. As for the roll wear, one can distinguish between (a) ordinary wear (i.e., the uniform loss of material over the roll surface by abrasion) increasing from early to later stands, and (b) surface defects, of which there are four basic types, with the severity increasing in the following order (Ryu et al., 1992):

1. Fire cracks, caused by thermal fatigue at the early stands, where strip temperature is highest and strip speed lowest, allowing longer contact time between the roll and hot strip. The cracks may be widespread (Figure 14.1a), but the oxide and roll shell surface do not fall off, even with deep cracks (Figure 14.1b).
2. Pitting, where small pieces of oxide or shell fall out, but the surface area of individual defects is less than 0.25 mm^2 (Boccalini and Sinatora, 2002).
3. Comet tails, that is, large pores at the roll surface (Figure 14.2), with the surface area of individual defects greater than 0.25 mm^2, around which larger chunks of oxide fall off, giving the worn area the shape of a comet tail (Uijtdebroeks et al., 1998).
4. Banding, where the significant pieces of shell and oxide are pulled out (Figure 14.3). Peterson (1956) attributed this to adhesive wear, however, Li, Xu, and He (2000) concluded that banding is caused by the formation and propagation of cracks along M_3C and M_7C_3 carbides, which suggests fatigue wear, which is consistent with de Barbadillo and Trozzi (1981).

14.2 Roll Surface Monitoring System Observations

Valuable information was obtained with an online surface monitoring system (Uijtdebroeks et al., 1998). The observations during two schedules in the Sidmar plant in Belgium are shown in Figure 14.4, and the following scenario was proposed:

1. The formation of oxide in the beginning of a schedule reduces friction.
2. Fire cracking and pitting follow, but they have no significant impact on COF. Fire cracks were not observed in schedule (b).
3. Formation of comet tails later in the schedule triggers a steady increase in COF.
4. COF reaches its maximum with banding.
5. Recovery of the oxide layer after banding reduces COF (Figure 14.4.b), but it remains high.

These observations suggest that fire cracks do not affect friction, perhaps because oxide is cracked, but not removed. Schröder (2003) claimed that fire cracks increase the COF, without evidence. As seen in Figure 14.4, comet tails increase the COF, presumably because oxide is worn to the extent that the

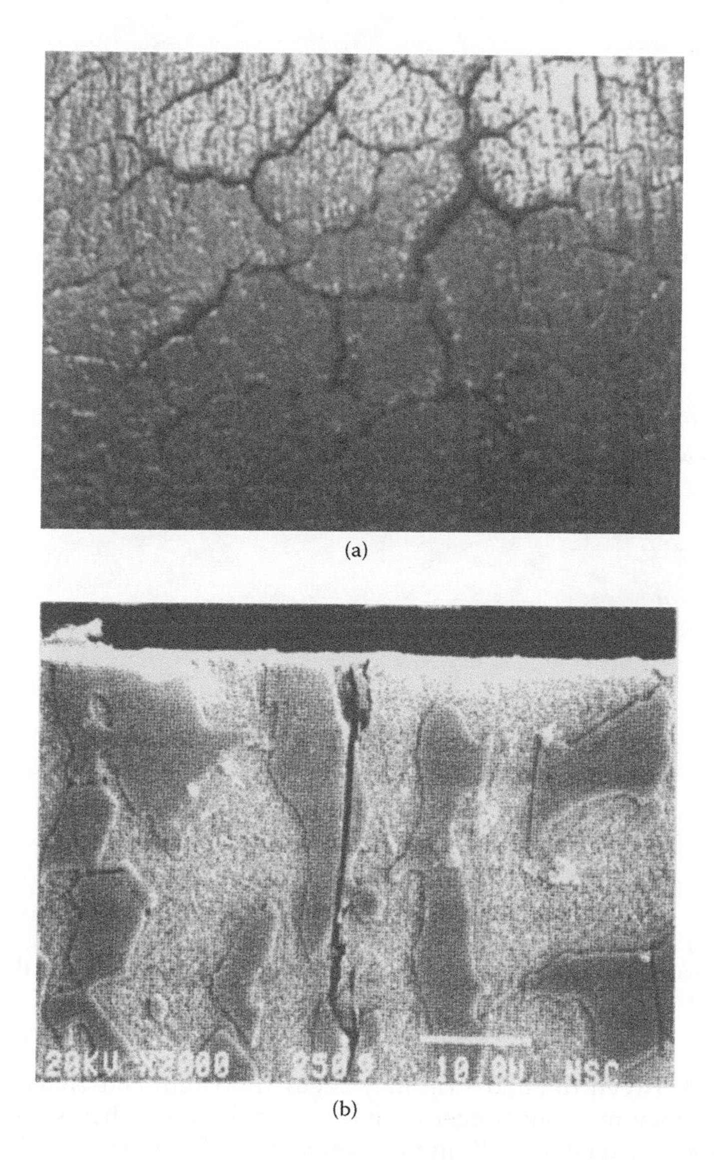

(a)

(b)

FIGURE 14.1
(a) Fire cracks on the roll surface. (Reprinted from F.J. Belzunce, A. Ziadi, and C. Rodriguez, *Eng. Failure Anal.* 11:789–797, 2004. With permission from Elsevier.) (b) Cross-section of a fire crack, perpendicular to roll surface. (Reprinted from O. Kato, H. Yamamoto, M. Ataka et al., *ISIJ Int.* 32:1216–1220, 1992. With permission from Iron and Steel Institute of Japan.)

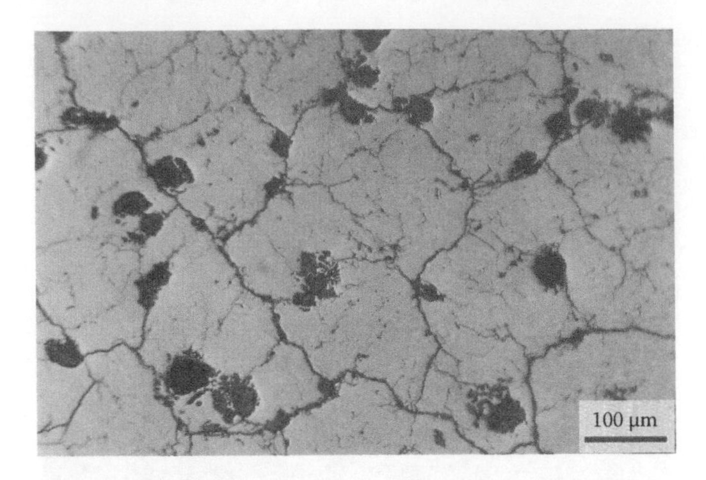

FIGURE 14.2
Comet tails. (Courtesy of Dr Mario Boccalini, Jr)

FIGURE 14.3
Banding. Roughness in the unaffected area is Ra = 0.95 μm, and Ra = 2.3 μm in the banding area. (Courtesy of Dr Mario Boccalini, Jr)

adhesion between roll and strip may occur (Boccalini and Sinatora, 2002). However, they may not trigger an increase in friction. They were observable by the naked eye on rolls in the commercial mill, accompanied with fire cracks, and there was no noticeable increase in friction in these schedules. This can be explained by the extent of areas affected by comet tails.

It is noteworthy that there were no significant instances of the friction changes after rolls were reused in the commercial mill, that is, inserted in stands without grinding after previous usage. The likely reason is that the roll oxide was still in reasonable condition, and friction would not be significantly affected if there were no peeling and pitting.

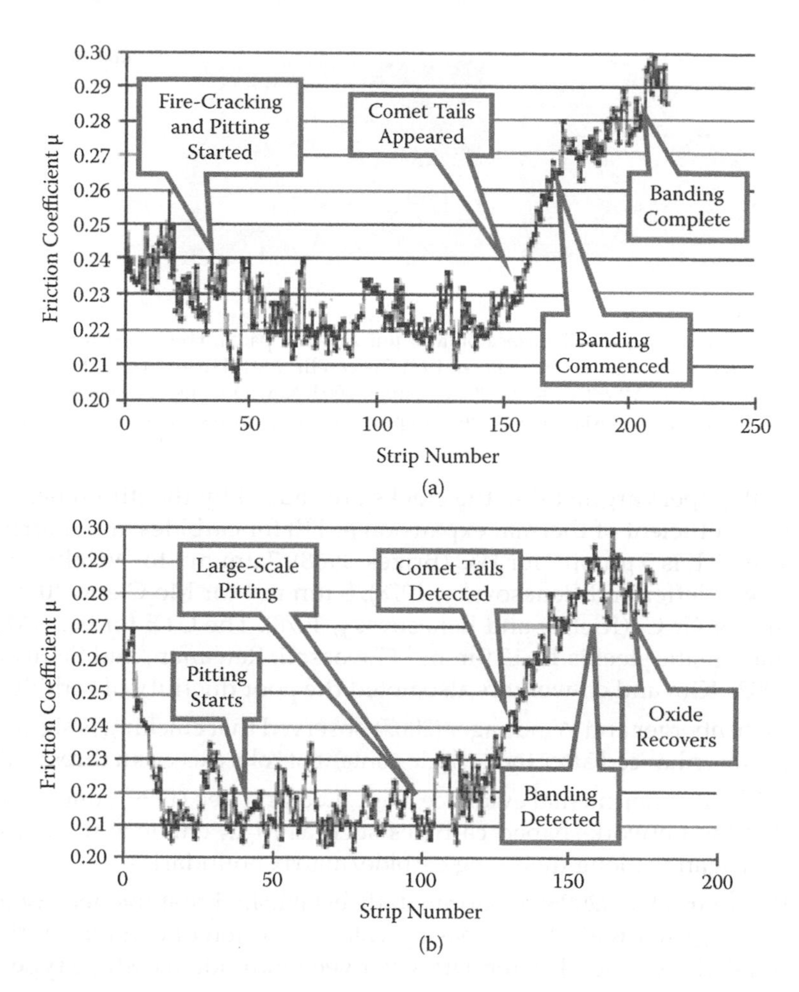

FIGURE 14.4
Evolution of wear and COF in schedules (a) and (b) with HiCr rolls. (Reprinted from H. Uijtdebroeks, R. Franssen, G. Sonck, et al., *La Rev. Met.* 95:789–799, 1998. With permission from EDP Sciences.)

14.3 Formation of Fire Cracks and Depth of Damage on Different Roll Types

There is a consensus that carbides are preferential sites for the initiation of fire cracks:

1. Mercado-Solis and Beynon (2005) claimed that cracks start around coarse carbides (i.e., MC, M_2C and M_7C_3), whereas Kim et al. (2004) observed the crack initiation around M_2C, M_6C and M_7C_3.

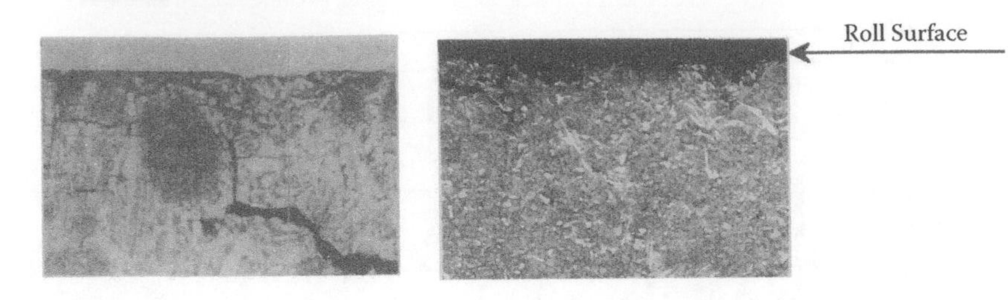

FIGURE 14.5
Cross section of damaged roll surface on a 70 μm × 50 μm patch, Hi-Cr cast iron (left) and HSS (right). (From Sorano, H., N. Oda, and J.P. Zuccarelli, 2004, History of high-speed steel rolls in Japan, in *Proc. MS&T Conf.*, 26–29 September 2004, New Orleans, 379–390, Warrendale: The Minerals, Metals and Materials Society. Reprinted with permission of the MS&T sponsor societies.)

2. Both papers argued that the cracks are caused by the difference in the coefficient of thermal expansion (CTE) for carbides and matrix. The CTE is 5 μm m^{-1} for WC (Brizes, 1968), 7 μm m^{-1} for VC (Brizes, 1968; Kieffer and Benesovsky, 1978), 8 μm m^{-1} for Mo_2C and 10 μm m^{-1} for Cr_3C_2 (Kieffer and Benesovsky, 1978). The CTE for 9Cr-1Mo martensitic steel is 11–13 μm m^{-1} (Tavassoli, Rensman, and Schirra, 2002). Kim and co-workers also noted the poor ductility of carbides.
3. Ziehenberger and Windhager (2007) observed that cracking is significantly enhanced after the carbide content of rolls exceeds a threshold.
4. Without experimental evidence, Park, Lee, and Lee (1999) contended that the network-shaped carbides such as M_7C_3 enhance the crack propagation along their long carbide/matrix boundaries.
5. Pellizzari et al. (2006) experimentally established that the density of cracks grows with the increasing volume portion of carbides in the roll shell, without differentiating between individual carbide types.

Sorano, Oda, and Zuccarelli (2004) observed that cracks propagate much deeper into HiCr rolls than HSS rolls (Figure 14.5), which agrees with the observations by Hashimoto et al. (1995). This could be associated with the higher total content of carbides in HiCr rolls.

References

Belzunce, F.J., A. Ziadi, and C. Rodriguez. 2004. Structural integrity of hot strip mill rolling rolls. *Eng. Failure Anal.* 11:789–797.

Boccalini, M. Jr, and A. Sinatora. 2002. Microstructure and wear resistance of high speed steels for rolling mill rolls. In *Proc. 6th Int. Tooling Conf., Karlstad, Vol. 1*, 425–438.

Brizes, W.F. 1968. *Mechanical Properties of the Group IVb and Vb Transition Metal Monocarbides*. Space Research Coordination Center, University of Pittsburgh.

De Barbadillo, J.J., and C.J. Trozzi. 1981. Mechanism of banding in hot strip mill work rolls. *Iron Steel Eng.* 58(1):62–71.

Hashimoto, M., T. Kawakami, T. Oda et al. 1995. Development and application of high-speed tool steel rolls in hot strip rolling. *Nippon Steel Tech. Rep.* No. 66:82–90.

Kato, O., H. Yamamoto, M. Ataka et al. 1992. Mechanisms of surface deterioration of roll for hot strip rolling. *ISIJ Int.* 32:1216–1220.

Kieffer, R., and F. Benesovsky. 1978. Carbides (industrial heavy-metal). In *Kirk-Othmer Encyclopaedia of Chemical Technology Vol. 4*, 476–535. New York: John Wiley & Sons.

Kim, C.K., J.I. Park, J.H. Ryu et al. 2004. Correlation of microstructure and thermal-fatigue properties of centrifugally cast high-speed steel rolls. *Met. Mat. Trans.* 35A:481–492.

Li, C.S., J.Z. Xu, X. He et al. 2000. Black oxide layer formation and banding in high chromium rolls. *Mat. Sci. Tech.* 16:501–505.

Mercado-Solis, R.D., and J.H. Beynon. 2005. Simulation of thermal fatigue in hot strip mill work rolls. *Scand. J. Met.* 34:175–191.

Park, J.W., J.C. Lee, and S. Lee. 1999. Composition, microstructure, hardness and wear properties of high-speed steel rolls. *Met. Mat. Trans. A* 30A:399–409.

Pellizzari, M., D. Cescato, A. Molinari et al. 2006. Laboratory testing aimed at the development of materials for hot rolls. In *ATS Steel Rolling Conf.* Paris.

Peterson, C.E. 1956. Cause and prevention of hot strip work roll banding. *Iron Steel Eng.* 33 (12):98–101.

Ryu, J.H., O. Kwon, P.J. Lee et al. 1992. Evaluation of the finishing roll surface deterioration at hot strip mill. *ISIJ Int.* 32:1221–1223.

Schröder, K.H. 2003. *A Basic Understanding of the Mechanics of Rolling Mill Rolls*. Tenneck: Eisenwerke Sulzau-Werfen.

Sorano, H., N. Oda, and J.P. Zuccarelli. 2004. History of high speed steel rolls in Japan. In *Proc. MS&T Conf.*, 26–29 September, *New Orleans*, 379–390. Warrendale, PA: Minerals, Metals and Materials Society.

Stachowiak, G., and A.W. Batchelor. 2005. *Engineering Tribology*, 3rd ed. Boston: Butterworth-Heinemann.

Tavassoli, A.-A.F., J.-W., Rensman, and M. Schirra. 2002. Materials design data for reduced activation martensitic steel type F82H. *Fusion Eng. Des.* 61–62:617–628.

Uijtdebroeks, H., R. Franssen, G. Sonck et al. 1998. On-line analysis of the work roll surface deterioration in a hot strip mill. *La Rev. Met.* 95:789–799.

Ziehenberger, H.H., and M. Windhager. 2007. State of the art work rolls for hot rolling flat products. In *Proc. CONAC 2007, 3rd Steel Industry Conf. and Expo.*, Monterrey, Mexico, 11–14 November.

15

Friction Evolution over Schedules and Campaigns

In the commercial mill, friction varies in rather consistent patterns over schedules, and varies over roll campaigns. These phenomena are analysed and discussed in this section, because an understanding of this variability can give more insight into friction in hot rolling.

15.1 Schedule

Numerous observations of the evolution of friction on HSS and HiCr rolls have been reported. There are significant differences in the friction observed for the same roll types. Regarding HSS rolls, Kang et al. (2001) observed on a rig a consistent decline of COF with increasing rolled length (Figure 15.1), which agreed well with the plant observations by Sanfilippo et al. (2002). On the other hand, Gotoh et al. (1998) noted that on a rig the COF initially decreases, and then increases before stabilising (Figure 15.2), and the pattern was similar to the one observed in a plant by Hashimoto et al. (1995). However, Steinier and co-workers (1999) spotted a very different pattern in another commercial mill, where the COF initially increased and then declined.

As for HiCr rolls, observations on rigs (Figures 15.1 and 15.2), and in plants (Hashimoto et al., 1995; Sanfilippo et al., 2002) indicated that COF, generally, decreases with the cumulative rolled length. However, observations by Steinier et al. (1999) suggested the opposite trend, as well as those shown in Figure 14.4.

These large discrepancies can be explained by different rolling conditions, in terms of force, speed, reduction, temperature, lubrication and cooling regime, and only the comparisons made under consistent conditions are meaningful. Morales, Sandoval, and Murillo (1999) gave an example where COF evolution differed substantially between all coils and the coils rolled at a specific reduction (Figure 15.3).

For a comparison, the evolution on the rolls in the commercial mill was analysed. Two measures of friction were used, namely COF calculated with a

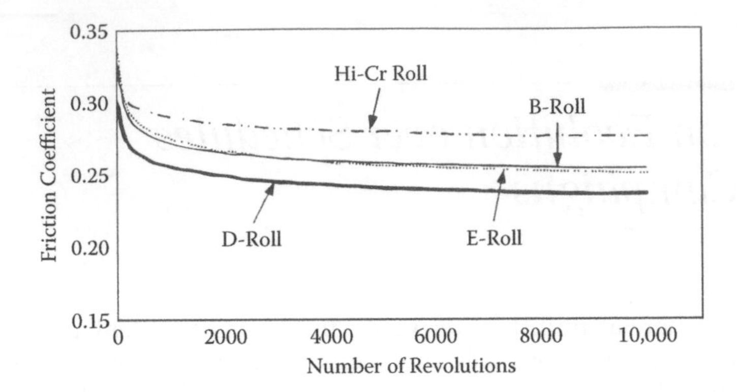

FIGURE 15.1
Evolution of the coefficient of friction over cumulative rolled length of HSS (B, D and E) and HiCr rolls. (Reprinted from Y.J., Kang, J.C. Oh, H.C. Lee et al. *Met. Mat. Trans. A* 32A:2515–2525, 2001. With permission from Springer Verlag.)

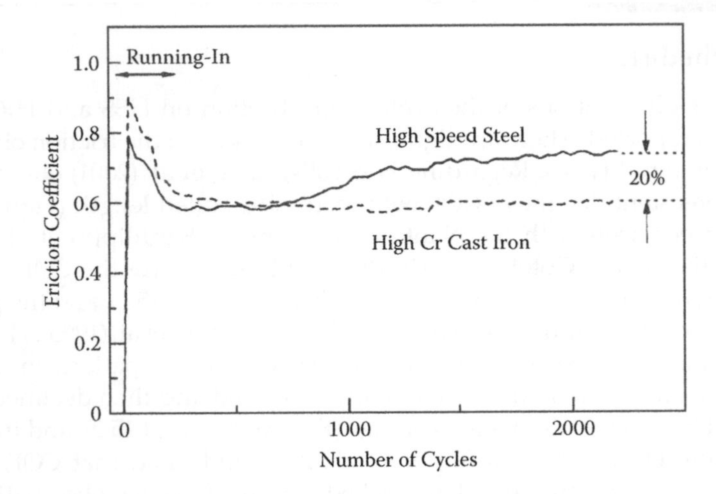

FIGURE 15.2
Evolution of the coefficient of friction over cumulative rolled length of HSS and HiCr rolls. (Reprinted from K. Gotoh, H. Okada, T. Sasaki et al. *Tetsu-to-Hagane* 84:861–867, 1998. With permission from Iron and Steel Society of Japan.)

proprietary formula (Section 17.2), and the force measured on the coils of the same grade and gauge, rolled under similar conditions. For both measures, the coils analysed were selected with the criteria shown in Appendix E. Both friction measures performed similarly, and only the COF trends are shown. The behaviour of HSS rolls at stands F1 and F2 can be summarised as follows:

1. Three datasets from F1 were analysed (Figure 15.4), and the trends in the first and the third set were similar, with COF first declining, then increasing. In the second set, COF generally declined over the schedule.
2. Most HSS rolls at F2 showed a steady decline of the COF over the schedule (Figure 15.5a). However, two pairs (advertised as low-friction rolls, Figures 15.5b and c) had a markedly different shape of COF evolution. Their COF was generally lower, but tended to increase towards the end of schedules. They generally differ from other HSS rolls by higher content of chromium (6wt-% versus 4 wt-%), Mo (8 wt-% versus 5 wt-%), silicon (0.95 wt-% versus 0.6 wt-%) and carbon (2.3 wt-% versus 2 wt-%).

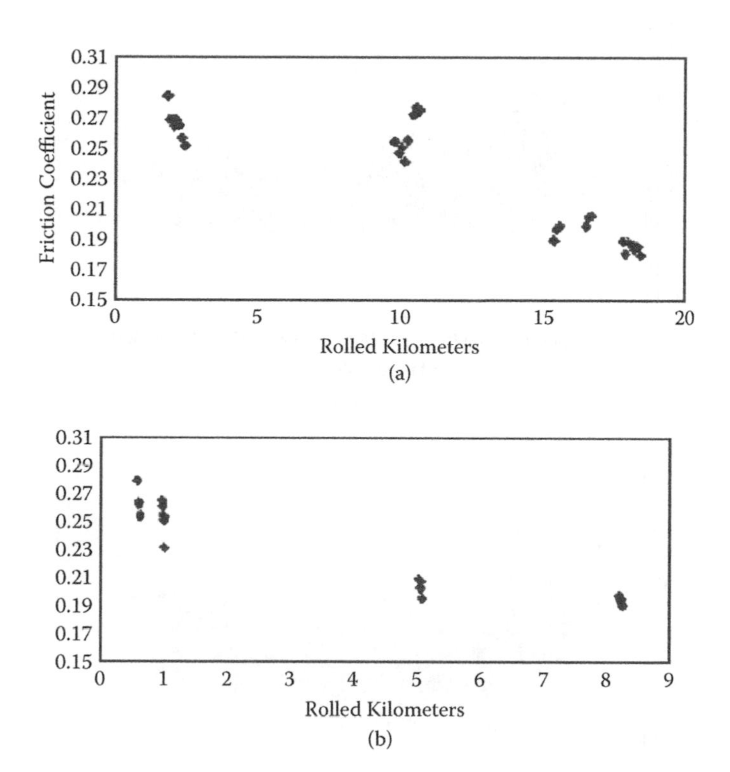

FIGURE 15.3
Evolution of COF on HiCr rolls at F1 for (a) all coils, and (b) those at specific reduction. (Reprinted from J. Morales, I. Sandoval, and G. Murillo, *AISE Steel Tech.* 76 (11):16 18, 1999. With permission from the Association for Iron and Steel Technology.)

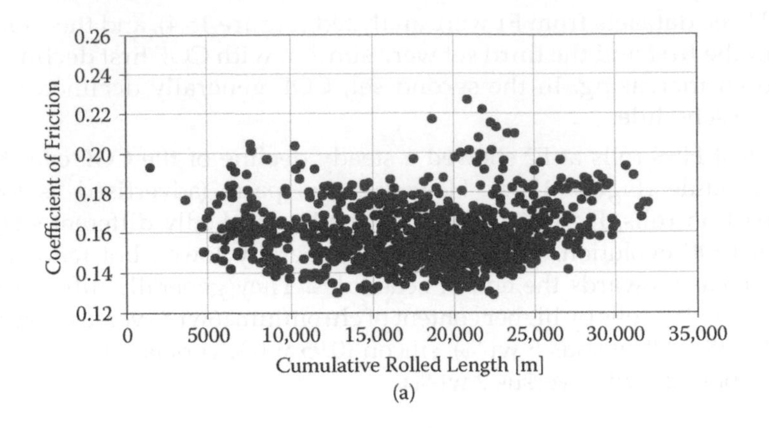

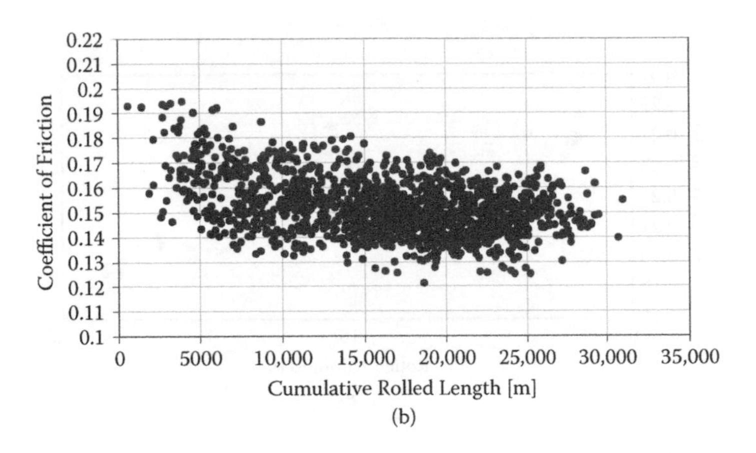

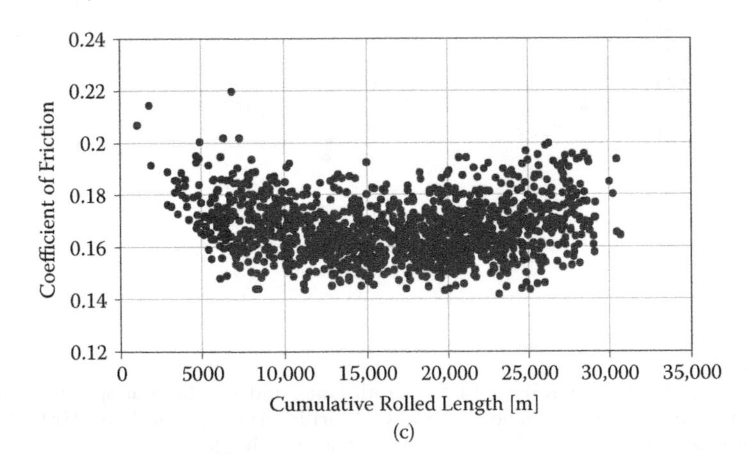

FIGURE 15.4
Evolution over schedule of coefficient of friction at stand F1. (a) Set 1; (b) Set 2; (c) Set 3.

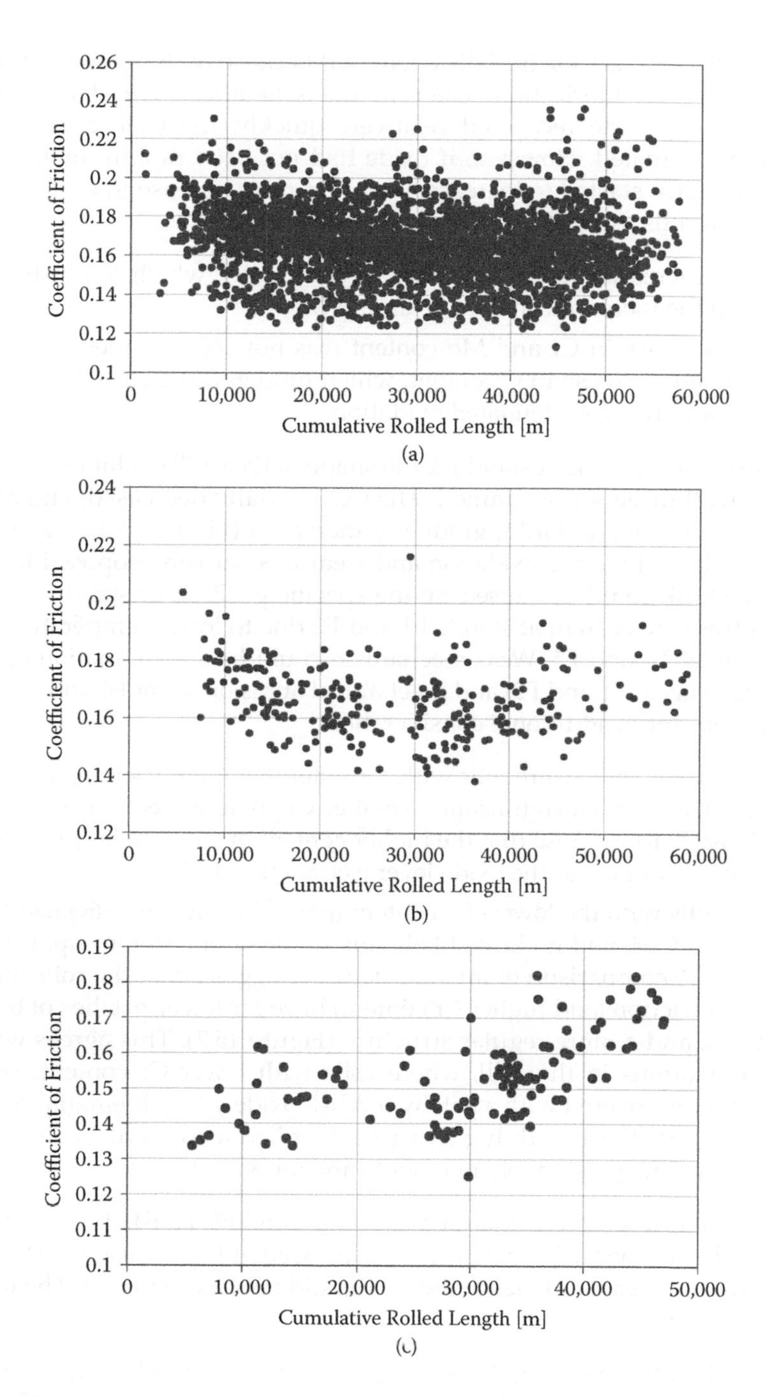

FIGURE 15.5
Evolution over schedule of coefficient of friction at stand F2. (a) All rolls save low-friction rolls; (b) low-friction pair 1; (c) low-fricton pair 2.

It is likely that the COF initially decreased because of the growth of oxide, which improved lubrication. Later in the schedule, wear damaged the oxide layer. If oxide recovered relatively quickly, the COF continued to decline or stagnated. However, if oxide had not been rapidly replenished, COF could have started to increase. Regarding the increase in COF of those low-friction rolls, either:

1. Wear increased faster than on the other rolls, which is not likely given the increased content of Cr and Mo.
2. The increase in Cr and Mo content was not accompanied with an adequate increase in C content, which hindered the carbide formation, and free Cr attenuated oxidation.

Regarding HiCr rolls at stand F3, the shape of the COF evolution was very similar in all three sets examined. The COF initially decreased, and after a distinct minimum, a "kink", gradually increased (Figure 15.6). Again, this can be explained via the oxidation and wear mechanism proposed for HSS rolls. The kink could be caused by the specific position of stand F3. Oxide growth was slower than at stands F1 and F2 due to lower temperature, and faster than at F4 and F5. Wear mechanism could be a combination of thermal fatigue (as at F1 and F2) and mechanical abrasion (as at F4 and F5). Two observations merit additional consideration:

1. In each set, there were rolls with consistently higher friction, marked CR1–CR5, with a significantly smaller Cr content (Section 16.1). It is plausible to suppose that the higher content of Cr carbides provided a better support to the oxide layer (see Section 11.5).
2. The rolls with the lowest Cr content (pair CR4, Figure 15.6c) had the highest COF and no kink, likely due to the poor carbide support to oxide. A comparison of microstructures suggests that the rolls with lower friction (and higher Cr content) have far fewer patches of bare metal, and a more regular structure (Figure 15.7). This agrees with observations in the mill, where rolls with lower Cr content were generally covered with a patchy reddish oxide, likely hæmatite, after extraction; those with higher Cr content had a contiguous cover of black oxide, presumably magnetite and/or wüstite.

Five ICDP roll sets were analysed at both F4 and F5. At F4, the evolution is illustrated with the COF. Specific force was used at F5 because the strip exit speed was not measured, hence the COF could not be calculated. The results are as follows:

1. At F4, in three sets (1, 2 and 4, Figure 15.8a), friction kept gradually increasing over the schedule. In the other two sets (Figure 15.8b), friction decreased slightly in the beginning of the schedule, and kept gradually increasing afterwards.

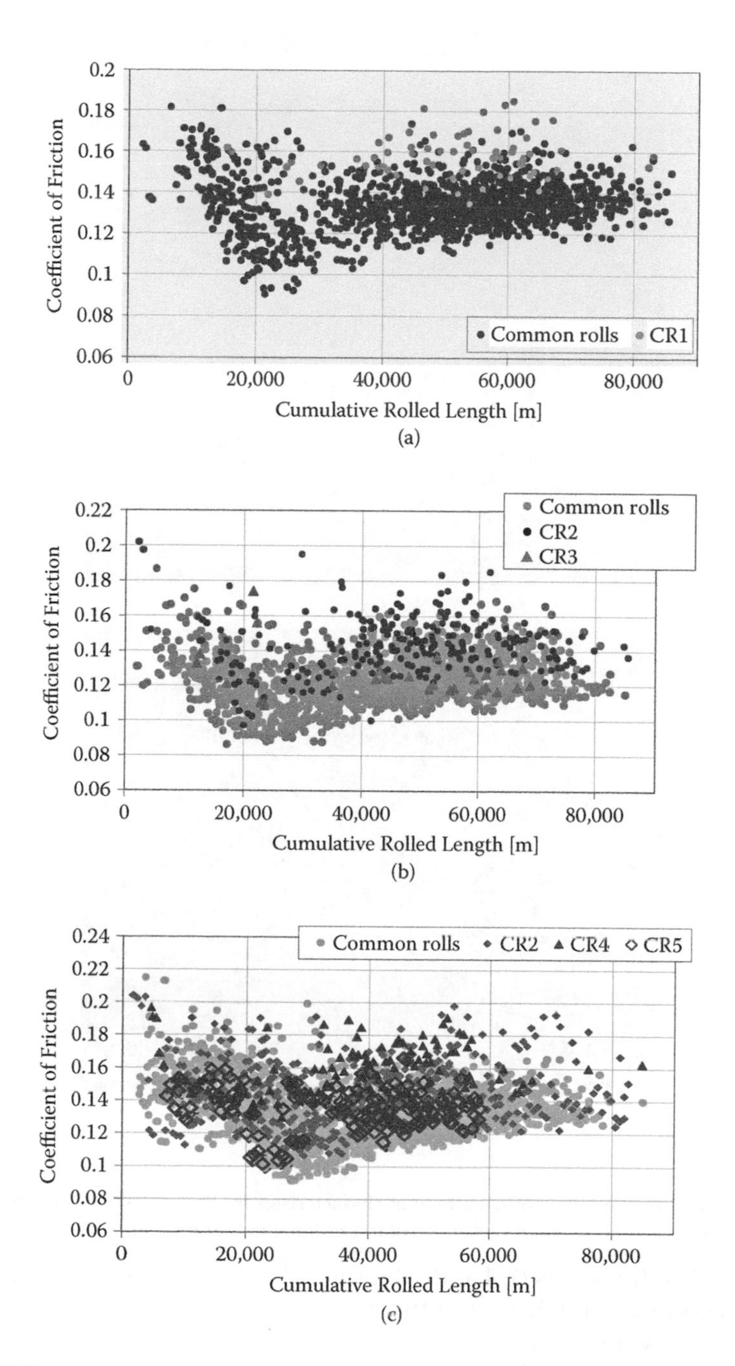

FIGURE 15.6
Evolution over schedule of coefficient of friction at stand F3. (a) Set 1; (b) set 2; (c) set 3.

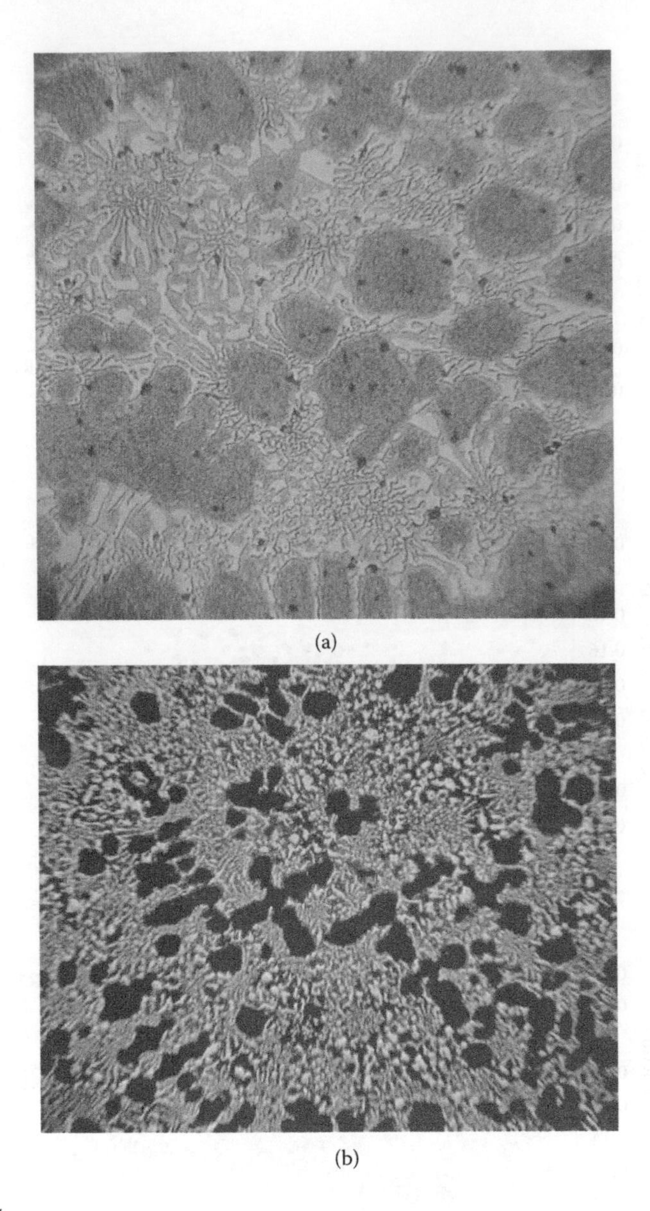

FIGURE 15.7
Microstructure of two HiCr rolls. Carbide is characterised by lighter colour. (a) 14.9 wt-% Cr; (b) 17.8 wt-% Cr. (From the private collection of the author. Photographs taken by Mr Boris Srkulj.)

2. At F5, in three sets (1, 4 and 5, Figure 15.9a), the friction measure was practically flat, whereas in the other two sets (Figure 15.9b), it kept gradually increasing.
3. The patterns could be explained by the oxidation and wear. The COF decreased when oxidation was more prominent than wear, and vice versa.

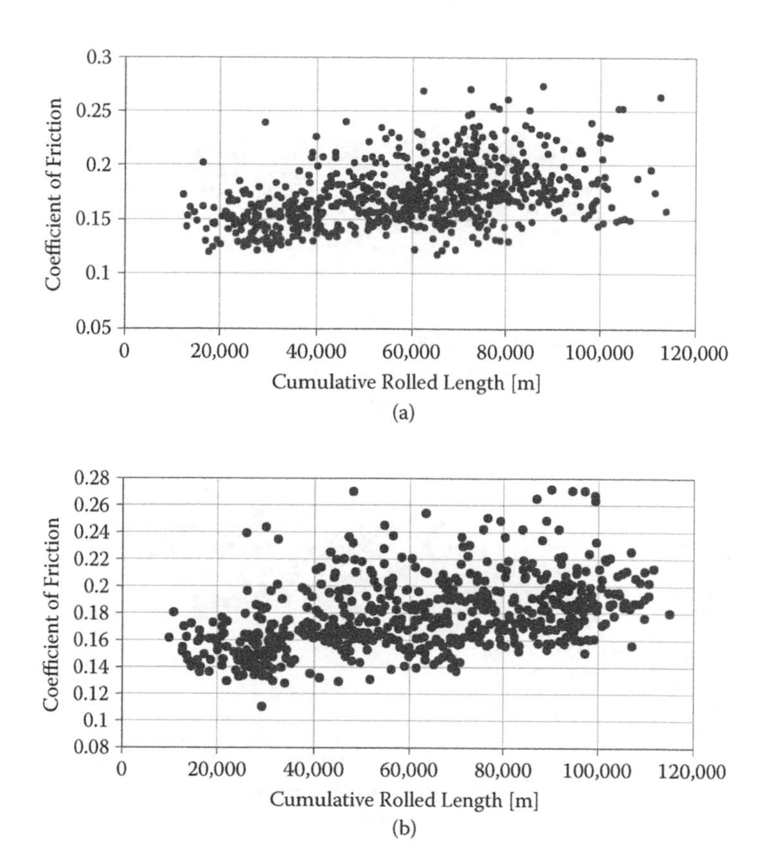

FIGURE 15.8
Evolution over schedule of coefficient of friction at stand F4. (a) Set 4; (b) set 5.

15.2 Campaign

Changes in the average COF over the campaign life of rolls were observed in the commercial mill. This issue was analysed with the coils selected by the criteria outlined in Appendix E. Only the rolls whose diameter changed by at least 50 mm per campaign were included:

1. *HSS.* There was no clear-cut trend with the first set at F1 (Figure 15.10a), although variability lessened at roll diameter <725 mm. The second set was more eventful. Save a "low-friction" pair, COF slightly decreased and then gradually increased as the campaign progressed (Figure 15.10b). With the 'low-friction' pair, the variability rapidly increased when the diameter decreased below

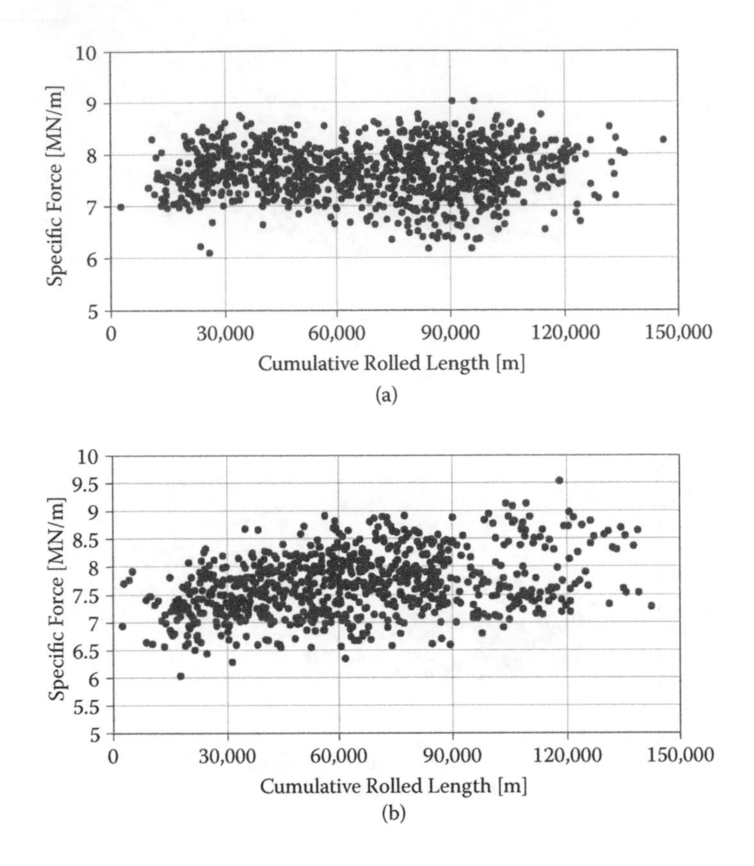

FIGURE 15.9
Evolution over schedule of coefficient of friction at stand F5. (a) Set 1; (b) set 3.

725 mm (Figure 15.10c). At F2, there was a significant increase in friction as the campaign progressed, followed by a stabilisation and an unexplained sudden drop at the 680 mm diameter (Figure 15.10d).

2. *HiCr.* These rolls changed trend often, particularly the second set (Figure 15.11).
3. *ICDP.* Friction steadily rose over the campaign for the first set at F4, but was variable with the second set (Figure 15.12). Figure 15.13 shows quite pronounced variability at F5.

15.3 Discussion

In Figures 15.4–15.6 the COF tended to decrease in the beginning of the schedules. An opinion is often heard that this was caused by the rapid polishing of the roll surface. However, the dependence of the roll surface roughness

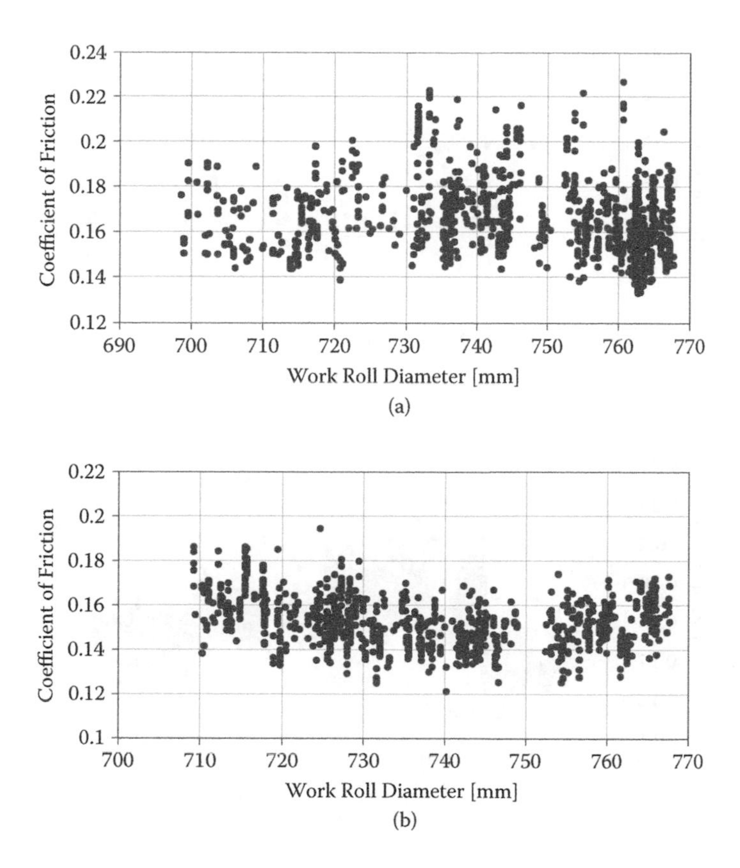

FIGURE 15.10
Friction evolution over campaign life of HSS rolls. (a) Set 1 at F1; (b) set 2 at F1 (without 'low-friction' pair). *(continued)*

on rolled length is unclear. Srkulj (2010) measured the roughness of HiCr rolls, after schedule, over the sections with and without contact with strip. The area with the contact had, on average, roughness 450% greater than the area without. Hence, roughness increases with the rolled length, but the friction of HiCr rolls tended to be largest at the start of schedules (Figure 15.6). On the other hand, the results obtained on twin-disc rigs differ widely (Gotoh et al., 1998; Lee et al., 2001; Kang et al., 2001; Pellizzari, Molinari, and Straffelini, 2005; Mercado-Solis and Beynon, 2005; Malbrancke, Uijtdebroeks, and Walmag, 2007).

Given the lack of a clear relationship between roll roughness and friction, the view of Uijtdebroeks et al. (1998) seems more plausible: that oxide formation reduces friction. The virtual absence of the drop in friction at stands F4 and F5 can be explained by the poorer roll oxidation at later stands, caused by the lower surface temperature. Relatively small differences in roll surface temperature can have a significant impact on the thickness of the oxide layer.

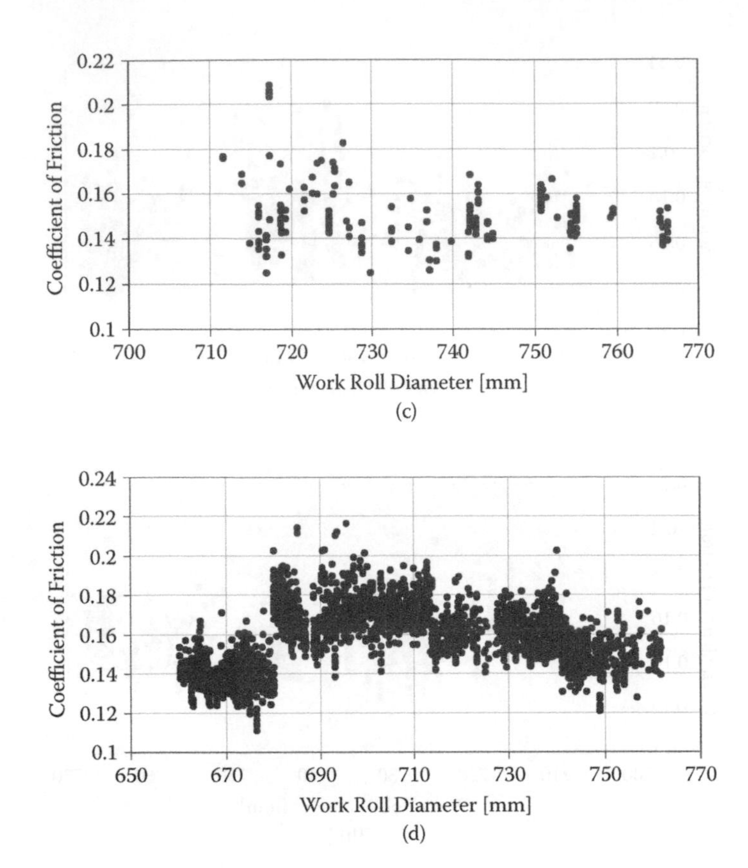

FIGURE 15.10 (continued)
Friction evolution over campaign life of HSS rolls. (c) set 2 at F1 (the 'low-friction' pair); (d) F2.

Tests by Zhu et al. (2010), with the oxidation of HSS specimens at temperatures of 550°C–700°C, showed that the space of 50°C can separate the cases of barely visible and clearly visible oxide. In the commercial mill, the difference between the surface temperatures at the early and the last stands easily exceeds 50°C.

An interesting issue here is the relationship between the roughness and the oxidation of rolls. The findings in laboratory investigations are contradictory. Gotoh et al. (1998) found that roughness increases with the oxide layer thickness for both HiCr and HSS rolls. Zhu et al. (2010) observed on HSS rolls that oxide roughness increases with the roll surface temperature. The increasing temperature enhanced oxide growth, hence roughness increased with oxidation. On the other hand, Pellizzari et al. (2005) contended that roughness is higher for poorly oxidized specimens (Figure 15.14).

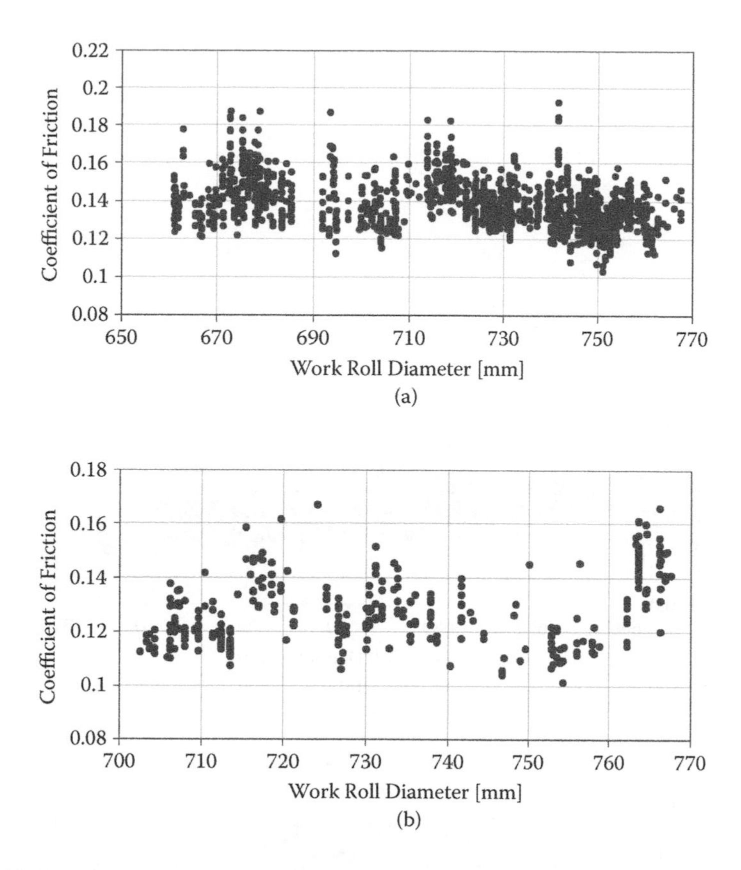

FIGURE 15.11
Friction evolution over campaign life of HiCr rolls at stand F3. (a) Set 1; (b) set 2.

Variability in the COF during campaigns is likely caused by radial inhomogeneity of shells. In centrifugally cast HSS rolls, high-density atom clusters move to the shell periphery due to centrifugal force, and lighter ones stay closer to the shell/core interface (Fu, Xiao, and Xing, 2008). W and Mo have density greater than liquid steel, and their concentration decreases away from the periphery (Figure 15.15). The opposite case is with species lighter than liquid iron (V and C), and there is no clear trend for Cr having a similar density to liquid steel. Hence, the outer shell contains more M_2C and M_6C carbides, and the inmost layer more MC carbides. Of course, centrifugal casting is not the only mechanism of roll manufacturing, and this variability was noted on the rolls produced differently. It is possible that the non-uniformity was also caused by poor temperature control during solidification and the roll surface treatment.

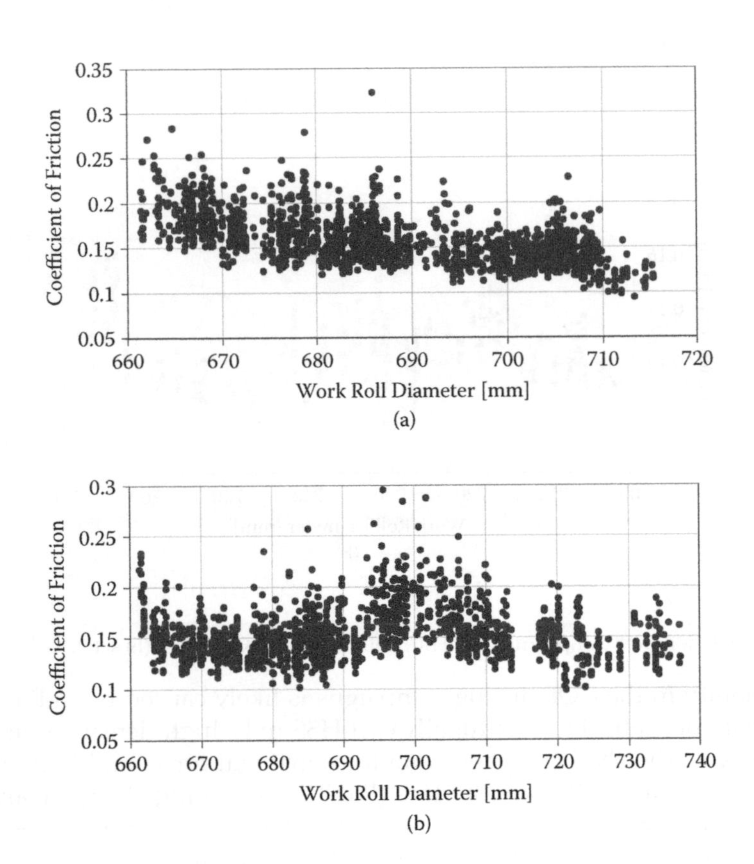

FIGURE 15.12
Friction evolution over campaign life of ICDP rolls at stand F4. (a) Set 1; (b) set 2.

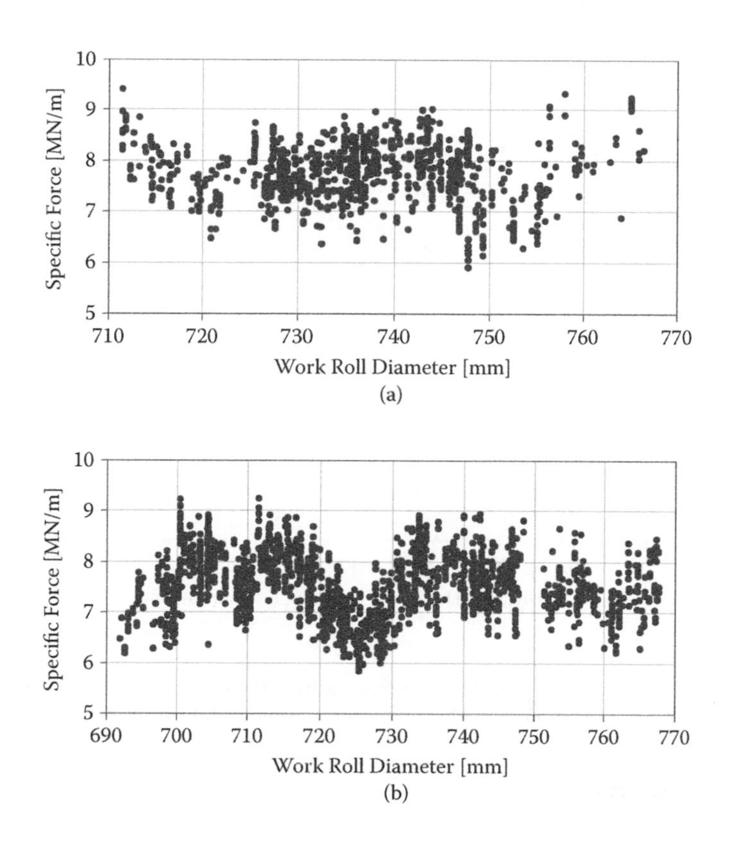

FIGURE 15.13
Friction evolution over campaign life of ICDP rolls at stand F5. (a) Set 1; (b) set 2.

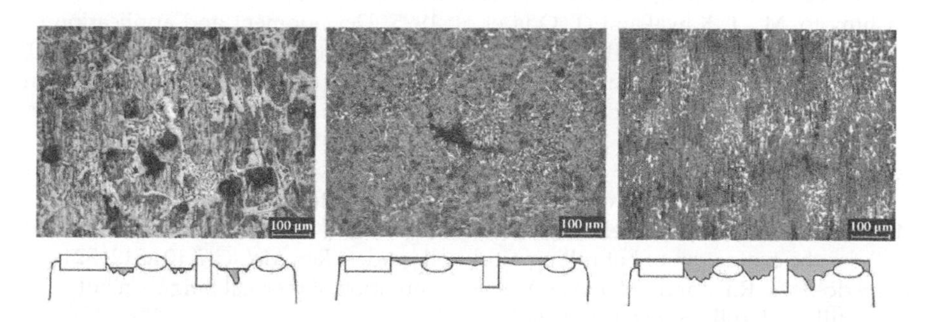

FIGURE 15.14
Relationship between roughness and oxidation. The surface appearance is shown at the top, and the oxide–roll interface at the bottom. The poor oxide layer formed on infinite chill iron with high Ra (left), the good oxide layer on HSS with low Ra (centre), and the 'optimum' oxide layer on HiCr with very low Ra (right). (Reprinted from M. Pellizzari, A. Molinari, and G. Straffelini, *Wear* 259:1281–1289, 2005. With permission from Elsevier.)

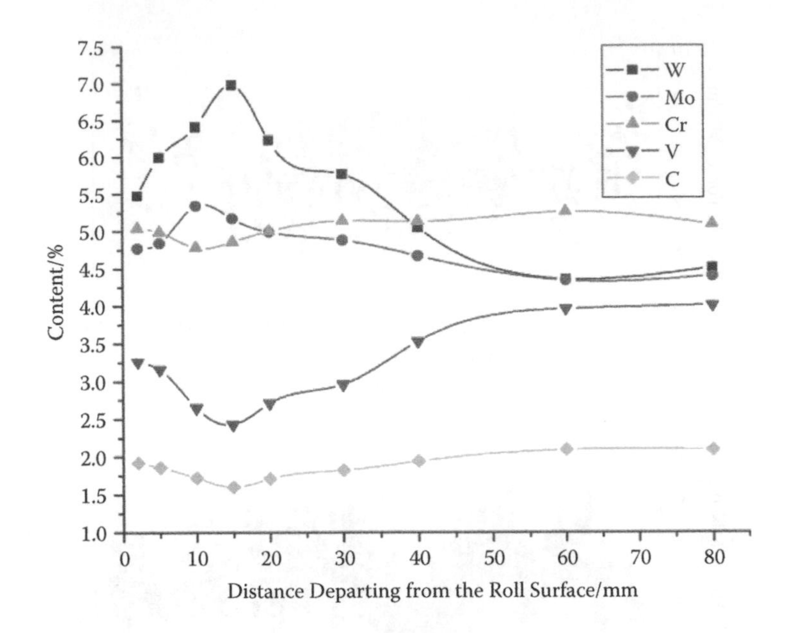

FIGURE 15.15
Distribution of key elements in roll shell along radius. (Reprinted from H. Fu, Q. Xiao, and J. Xing, *Mat. Sci. Eng. A* 479:253–260, 2008. With permission from Elsevier.)

References

Fu, H., Q. Xiao, and J. Xing. 2008. A study of segregation mechanism in centrifugal cast high speed steel rolls. *Mat. Sci. Eng. A* 479:253–260.

Gotoh, K., H. Okada, T. Sasaki et al. 1998. Effects of roll surface deteriorations on scale defect in hot rolling. *Tetsu-to-Hagane* 84:861–867.

Hashimoto, M., T. Kawakami, T. Oda et al. 1995. Development and application of high-speed tool steel rolls in hot strip rolling. *Nippon Steel Tech. Rep.* No. 66:82–90.

Kang, Y.J., J.C. Oh, H.C. Lee et al. 2001. Effects of carbon and chromium additions on the wear resistance and surface roughness of cast high-speed steel rolls. *Met. Mat. Trans. A* 32A:2515–2525.

Lee, J.H., J.C. Oh, J.W. Park et al. 2001. Effects of tempering temperature on wear resistance and surface roughness of a high speed steel roll. *ISIJ Int.* 41:859–865.

Malbrancke, J., H. Uijtdebroeks, and G. Walmag. 2007. A new breakthrough method for the evaluation of hot rolling work roll grades. *Rev. Met.-CIT* 104:512–521.

Mercado-Solis, R.D., and J.H. Beynon. 2005. Simulation of thermal fatigue in hot strip mill work rolls. *Scand. J. Met.* 34:175–191.

Morales, J., I. Sandoval, and G. Murillo. 1999. Influence of process parameters on friction coefficient of high-chromium rolls. *AISE Steel Tech.* 76 (11):46–48.

Pellizzari, M., A. Molinari, and G. Straffelini. 2005. Tribological behaviour of hot rolling rolls. *Wear* 259:1281–1289.

Sanfilippo, F., V. Lanteri, F. Geffraye et al. 2002. *Lubrication in hot Rolling, Effect of Different Utilisation Strategies on Strip Quality and Process Conditions for Various Steel Grades*. Eur. Com., Rep. EUR 20208EN.

Srkulj, B. 2010. Private communication.

Steinier, D., D. Liquet, J. Lacroix et al. 1999. Effect of processing parameters in the front stands of a HSM on the performance of HSS work rolls. In *Proc. 41th MWSP Conf.*, ISS.

Uijtdebroeks, H., R. Franssen, G. Sonck et al. 1998. On-line analysis of the work roll surface deterioration in a hot strip mill. *La Rev. Met.* 95:789–799.

Zhu, Q., H.T. Zhu, A.K. Tieu et al. 2010. In-situ investigation of oxidation behaviour in high-speed steel roll material under dry and humid atmospheres. *Corrosion Sci.* 52:2707–2715.

16

Relationship between Friction and Chemical Composition of Rolls

Analyses of the data from the commercial mill showed large differences in friction measures between rolls of the same type, but with different composition of roll shells. In this section, these differences are shown first, followed by the analysis of the relationship between friction and roll chemistry*, and discussions.

16.1 Differences in Friction between Same Type Rolls

The differences between the roll pairs of the same type are shown in Figure 16.1. Although the differences between HSS and HiCr roll pairs were clearly associated with the roll manufacturers, there were significant differences between the ICDP rolls of the same brand. It is interesting that pairs B1–B3 of HSS rolls (Figure 16.1b) contained 1.1 wt-% of cobalt in shells and that its addition alone was ineffective in reducing friction (*cf.* Section 12.5).

16.2 Relationship between Friction and Chemistry of Roll Shells

This relationship was investigated using multiple linear regression (MLR), with the species considered C, Cr, Mo, Mn, Ni, Si, P and S. Interestingly, no significant relationship was found between the friction and chemistry of ICDP rolls. On the other hand, the level of significance of MLR was below 1.2×10^{-3} for HSS and HiCr rolls. One cannot apply these formulae to predict quantitatively the COF of a single roll pair, which may be influenced by mill conditions, speed, force, temperature or grade. However, when several pairs of different composition are considered, the formulae can be expected to predict their relative friction with acceptable accuracy.

* The criteria for the data selection are summarised in Appendix E.

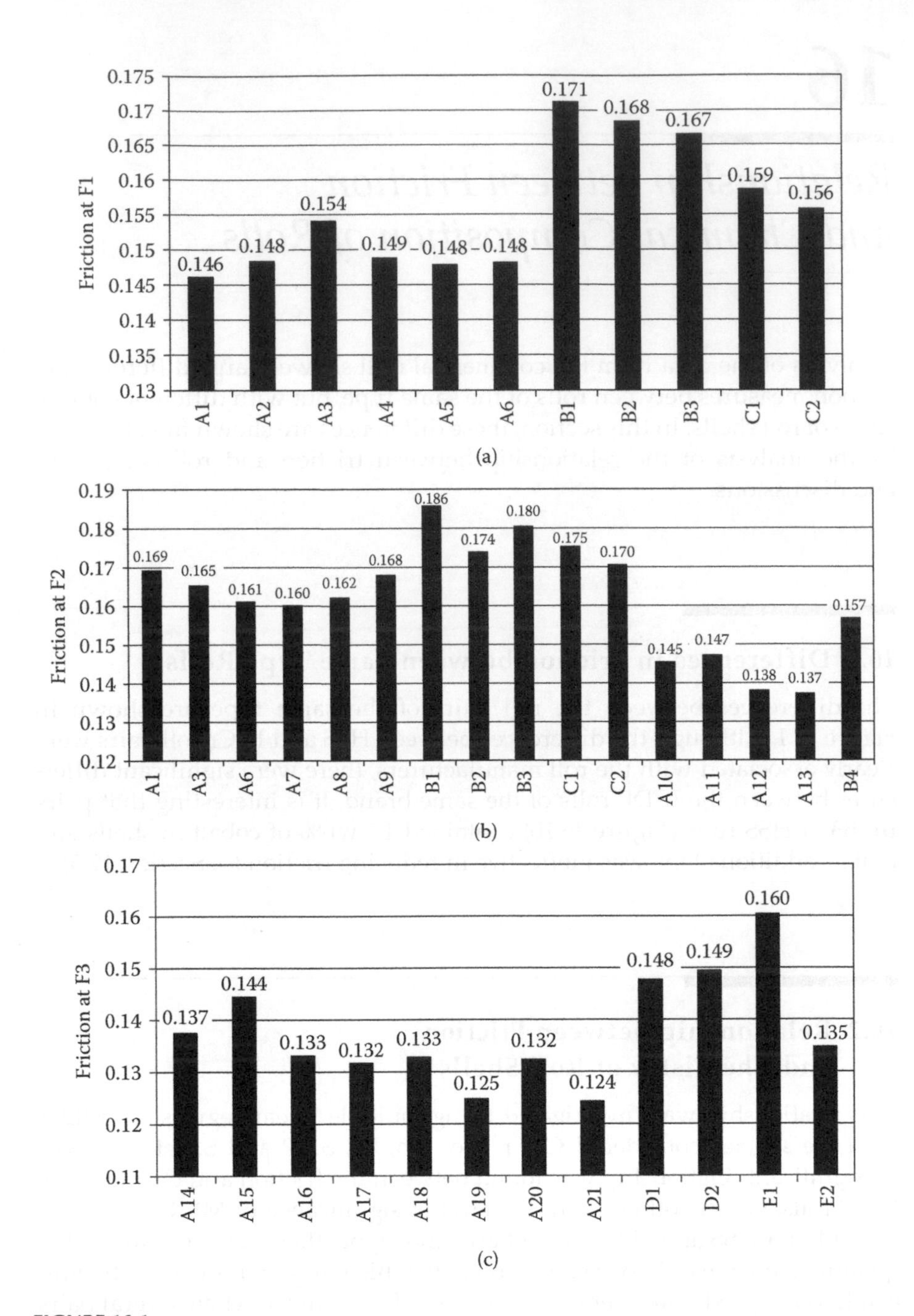

FIGURE 16.1
Friction at F1–F4 and specific force at F5 for different roll pairs. Letters A–E denote different roll manufacturers. The statistical significance of the difference between individual roll pairs at the same stand is $\alpha < 0.01$. (a) HSS at F1; (b) HSS at F2; (c) HiCr at F3; (d) ICDP at F4; (e) ICDP at F5.

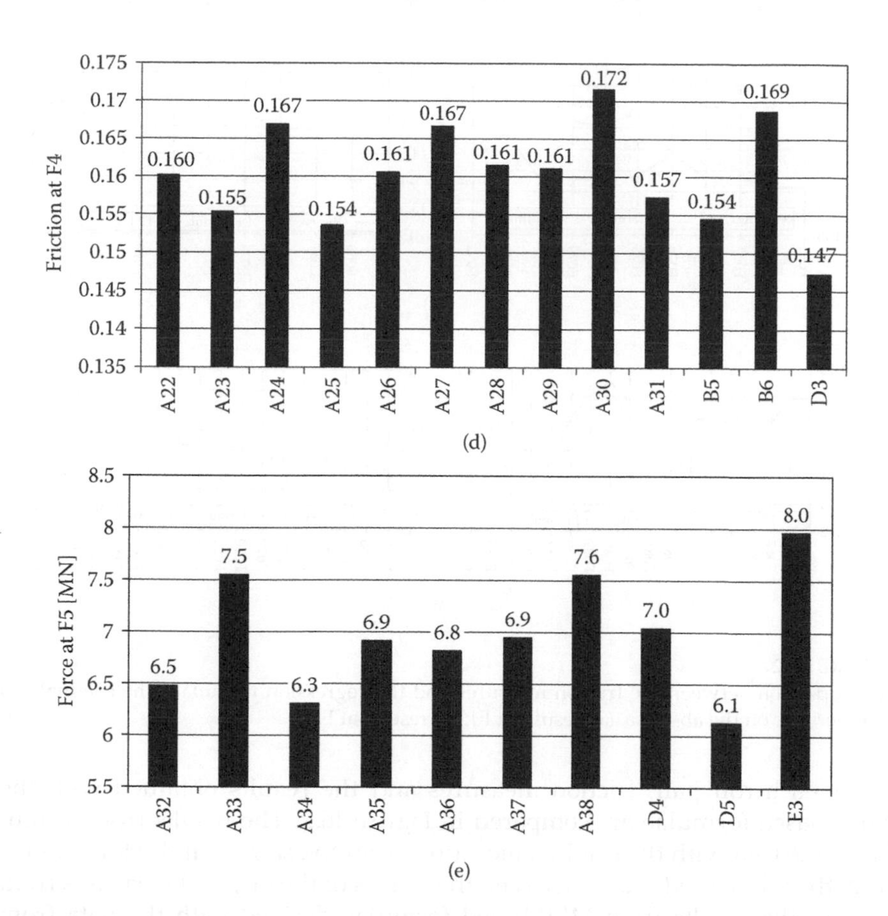

FIGURE 16.1 (continued)
Friction at F1–F4 and specific force at F5 for different roll pairs. Letters A–E denote different roll manufacturers. The statistical significance of the difference between individual roll pairs at the same stand is $\alpha < 0.01$. (a) HSS at F1; (b) HSS at F2; (c) HiCr at F3; (d) ICDP at F4; (e) ICDP at F5.

Regarding HSS rolls, the formulae for the COF derived with the data from stands F1 and F2 are given in Equations (16.1) and (16.2), respectively. Formulae derived from the force measure at F1 and F2 are given in Equations (16.3) and (16.4), respectively.

$$\text{Friction} = 0.415 - 0.22 \times C + 0.051 \times Cr - 0.048 \times Si - 0.0024 \times Mo \tag{16.1}$$

$$\text{Friction} = 0.414 - 0.209 \times C + 0.051 \times C - 0.0105 \times Mo \tag{16.2}$$

$$\text{Force} = 31.7 - 13.17 \times C + 3.16 \times Cr - 2.72 \times Si - 0.29 \times Mo - 1.29 \times Mn \tag{16.3}$$

$$\text{Force} = 29.6 - 11.6 \times C + 2.36 \times Cr - 3.4 \times Si \tag{16.4}$$

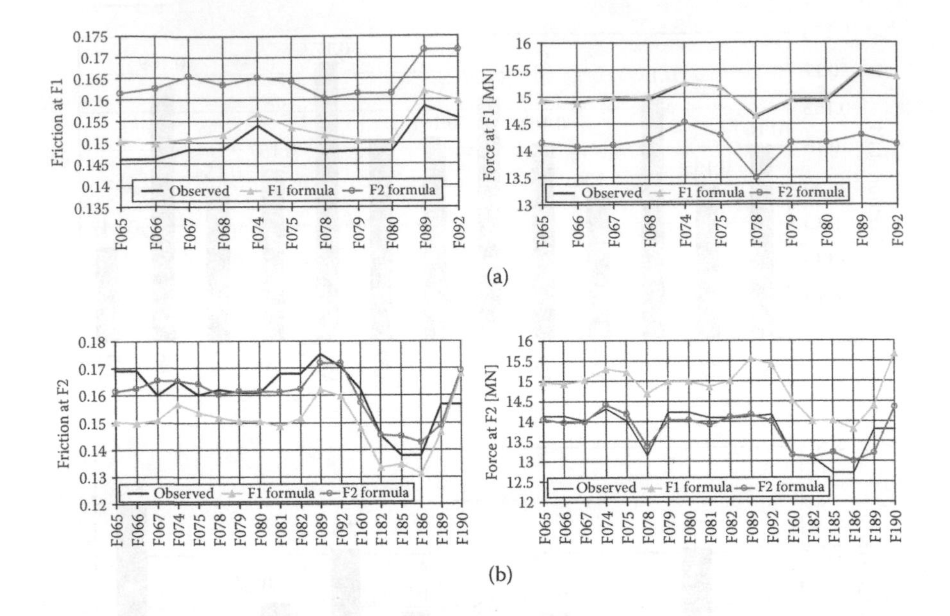

FIGURE 16.2
A comparison between the friction measures and the regression formulae. The original roll numbers are on the abscissa. (a) Results at F1; (b) results at F2.

For each roll pair, friction measures and the results obtained with the MLR-based formulae are compared in Figure 16.2. The results from formulae agree well with the original measures from the same stand. More important, there is a good match between the trends of the original measures from F1 and the results from MLR-based formulae derived with the data from stand F2, and vice versa.

Regarding HiCr rolls, three datasets were used. The equations for the COF are given by Equations (16.5)–(16.7), respectively, and for the force measure by Equations (16.8)–(16.10), respectively. For each set, the observed friction was compared to the results from the regression-based formulae derived from all three sets (Figure 16.3):

1. The formulae based on set 3 seem most suitable, inasmuch as they are based on the largest number of rolls. They would correctly anticipate high friction of F131 and F132 rolls in set 1. They would set a false alarm for F895, however.
2. Equations from sets 1 and 2 underestimated the friction measures of rolls F211/F212, because their composition is very different from those in the two sets.

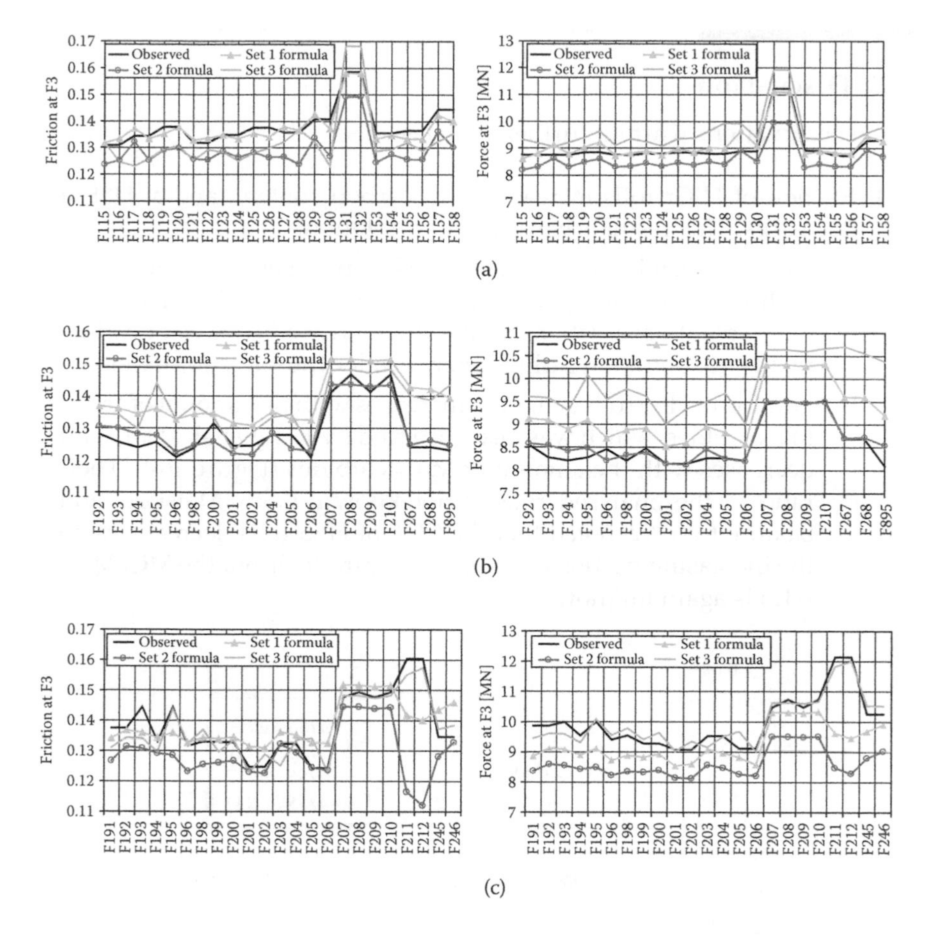

FIGURE 16.3
A comparison between the friction measures and the regression formulae. The original roll numbers are on the abscissa. (a) Results for set 1; (b) results for set 2; (c) results for set 3.

$$\text{Friction} = 0.173 + 0.0307 \times C - 0.00484 \times Cr - 0.0267 \times Mo \tag{16.5}$$

$$\text{Friction} = 0.047 + 0.0767 \times C - 0.00525 \times Cr - 0.0308 \times Mo \tag{16.6}$$

$$\text{Friction} = 0.3 - 0.017 \times C - 0.012 \times Cr + 0.016 \times Mo + 0.055 \times Ni \tag{16.7}$$

$$\text{Force} = 12.6 + 3.05 \times C - 0.57 \times Cr - 1.49 \times Mo \tag{16.8}$$

$$\text{Force} = 8.25 + 3.15 \times C - 0.35 \times Cr - 1.75 \times Mo \tag{16.9}$$

$$\text{Force} = 24.9 - 1.91 \times C - 0.77 \times Cr + 0.837 \times Mo + 2 \times Ni \tag{16.10}$$

16.3 Discussion

Three observations warrant further analysis:

1. Friction of HiCr rolls increases with carbon content, and at HSS it decreases. This result is hard for interpretation:
   a. It can be seen in Figure 16.4 that friction of HiCr rolls decreases with the increasing portion of carbides. A possible explanation of the plant observations is that the increased content of carbon that is not bound in carbides may hinder the oxide formation.
   b. Friction of HSS rolls generally increases with the amount of sharp MC carbides, which was also observed by Kang et al. (2001), and decreases with the amount of M_7C_3 carbides (Figure 16.4). Their ratio in the analysed rolls is not known. Goto, Matsuda, and Sakamoto (1992) noted that the COF of HSS rolls decreases with the increasing portion of carbides (Figure 16.5), but the MC/M_7C_3 ratio is again unknown.

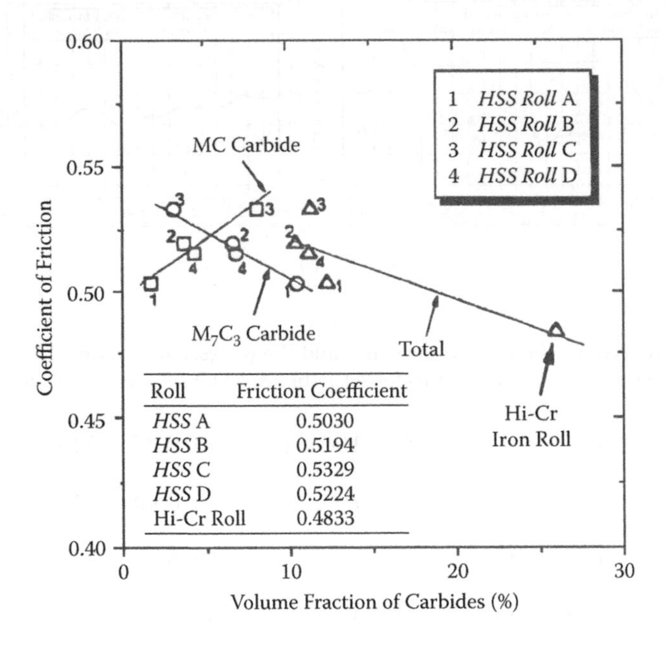

FIGURE 16.4
Effect on the COF of HSS samples of total carbide, MC and M_7C_3 content. (Reprinted from J.W. Park, J.C. Lee, and S. Lee, *Met. Mat. Trans. A* 30A:399–409, 1999. With permission from Springer Verlag.)

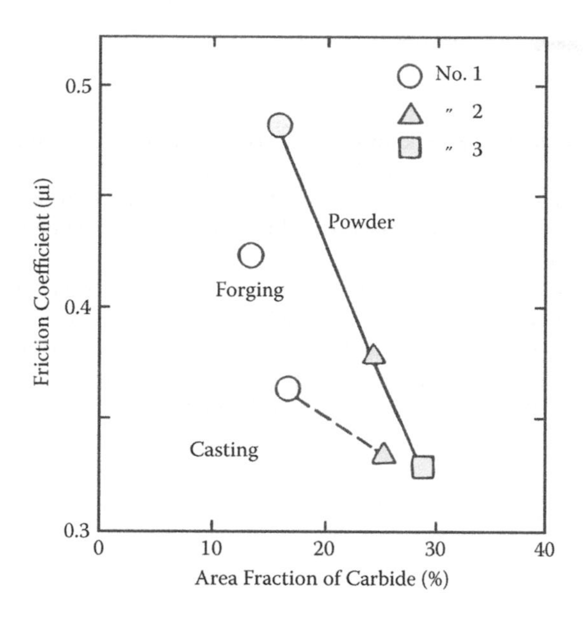

FIGURE 16.5
Effect of carbide content on the COF of HSS specimens. (Reprinted from K. Goto, Y. Matsuda, K. Sakamoto et al., *ISIJ Int.* 32:1184–1189, 1992. With permission from Iron and Steel Institute of Japan.)

2. At HiCr, friction decreases with Cr content, and at HSS it increases. As shown before, high Cr content may encourage the formation of carbides that could support oxide at HiCr rolls. The result for HSS rolls is hard to explain, because Figure 16.4 shows that friction of HSS rolls decreases with the fraction of M_7C_3 carbides, which are mainly Cr-based. Kang et al. (2001) also observed that friction of HSS rolls decreases with increasing Cr content. It is possible that in some cases carbon in HSS rolls preferentially forms MC carbides, and free Cr hinders oxygen formation.
3. At both HSS and HiCr rolls, friction decreases with Mo. Scandian et al. (2009) experimented with Mo additions to HiCr samples, although Cr content was higher than in the rolls analysed here (24–32 wt-% versus 12–19 wt-%). Molybdenum addition could reduce friction:
   a. Friction was lower for the composition 24 wt-% Cr – 3 wt-% Mo than for the Mo-free sample with 28 wt-% Cr.
   b. Friction of the 32 wt-% Cr sample progressively decreased when Mo content was changed from 0 to 6 wt-%, but increased at 9 wt-%.

References

Goto, K., Y. Matsuda, and K. Sakamoto. 1992. Basic characteristics and microstructure of high-carbon high speed steel rolls for hot rolling mill. *ISIJ Int.* 32:1184–1189.

Kang, Y.J., J.C. Oh, H.C. Lee et al. 2001. Effects of carbon and chromium additions on the wear resistance and surface roughness of cast high-speed steel rolls. *Met. Mat. Trans. A* 32A:2515–2525.

Park, J.W., J.C. Lee, and S. Lee. 1999. Composition, microstructure, hardness and wear properties of high-speed steel rolls. *Met. Mat. Trans. A* 30A:399–409.

Scandian, C., C. Boher, J.D.B. de Mello et al. 2009. Effect of molybdenum and chromium contents in sliding wear of high-chromium white cast iron: The relationship between microstructure and wear. *Wear* 267:401–408.

17

Mathematical Models of Friction in Steel Rolling

Although various attempts were made to develop a sensor for the measurement of friction in a mill, they have not produced a viable instrument as yet (Legrand, Lavalard, and Martins, 2012). Therefore, mathematical modeling is the only way to estimate friction in an industrial hot strip mill, essentially by calculating it from rolling force or torque. The models in general can be classified as empirical and first-principles based. These empirical, black-box models are reviewed first. Those employing first principles are analysed next, followed by a practical model for hot strip rolling based on the analyses of plant data from the commercial mill.

17.1 Empirical Models

Most empirical models use regression to develop formulae linking COF with speed, force and reduction; the others employ a correlation with temperature. There are ample reports in the literature on the links between friction and various individual operation parameters in hot rolling, but the strength of the correlation is uncertain:

1. *Velocity.* Researchers have tended to analyse the relationship with roll velocity, rather than slip, although it is the relative velocity between strip and roll that is more closely related to friction. Lee, Kwak, and Park (1996) observed a poor correlation in cold rolling, whereas Munther and Lenard (1997) reported that friction in hot rolling decreases with roll speed, other parameters being constant. On a 6-stand mill, Morales, Sandoval, and Murillo (1999) established that COF clearly decreases with roll velocity at F2 and F3, without a clear correlation at other stands. Finally, Sato et al. (1991) observed a complex relationship between sliding velocity and friction in a hot strip mill. At stand F2, the COF tended to decrease with this velocity, whereas at F5 it initially increased, and then decreased with it. At F7, the dependence showed almost a sinusoidal pattern.

2. *Temperature.* Koseki, Yoshida, and Inoue (1994) observed that the COF decreases with strip temperature, contrary to Sato et al. (1991). On the other hand, Inoue et al. (2002) claimed that the COF initially increases, then decreases; Munther and Lenard (1997) reported that the COF decreases with strip temperature for thin scale on strip (<290 μm), but increases with thick scale (1.59 mm). Milan et al. (2005) and Vergne et al. (2006) observed that the COF decreases with temperature for HSS rolls, and for HiCr it decreases up to 600–650°C, increasing at higher temperatures.
3. *Reduction.* Koseki et al. (1994) and Munther and Lenard (1997) reported that friction increases with thickness reduction if other parameters are constant, and Morales et al. (1999) observed in a 6-stand HSM that COF decreases with reduction.
4. *Scale thickness.* The differences between numerous observations are discussed in Chapter 13.3.1.
5. *Strip chemistry.* Sato et al. (1996) noted that the COF decreases with Si strip content. However, an analysis of 18 months worth of data from the commercial mill showed that the COF was distinctly highest for the grade with the largest Si content. No link could be established between COF and other significant species found in strip.

Regarding the relationship between the COF and multiple parameters, several models were devised using data fitting. The first three models were not tested with the mill data because they considered unspecified constants, and the fourth one performed poorly:

1. Koseki et al. (1994), for the strip temperature range of 850–1100°C:

$$\mu = dr^n\left(1+b/H_S\right) \tag{17.1}$$

   where H_S is scale thickness, and d, b and n are unspecified constants;
2. Lee et al. (1996):

$$\mu = C_0\exp\left(-C_V v_R - C_L L\right) \tag{17.2}$$

   where L is the cumulative rolled length, and C_o, C_v and C_L are unspecified constants;
3. Sato et al. (1991):

$$\mu = e_0 + e_1 x_{Si}^2 + e_2 x_{Si} + e_3 T_S^2 + e_4 T_S + e_5 \Delta v^3 + e_6 \Delta v^2 + e_7 \Delta v \tag{17.3}$$

   where x_{Si} is the wt-% of Si in strip, and T_S the strip temperature; and

4. Lenard and Barbulovic-Nad (2002):

$$\mu = 0.363 p/\sigma - 0.36 \tag{17.4}$$

where p is the rolling pressure, and σ flow stress.

Ginzburg and Ballas (2000) listed a variety of empirical models, mainly based on temperature, although a few included speed and thickness reduction as well. Some of them were based on the data collected during normal operation; the others used the information obtained at skidding. Generally, their performance with the plant data was not satisfactory.

The mill data from stands F1–F4 were then used to investigate the link with operational parameters systematically, under the following conditions:

1. COF was calculated with the model in Equation (17.7).
2. Only the rolls with similar friction (COF difference within ±0.01 with respect to Figure 16.1) and the coils rolled under similar conditions (Appendix E) were included.
3. Parameters were investigated one at a time. For reduction, the constraint on entry thickness was removed. Temperature dependence was tested with the entry and exit mill temperature.
4. Speed was represented in four ways, as roll speed v_R, strip speed v_S, speed differential $\Delta v = v_S - v_R$, and slip [Equation (17.5)]. Reduction was calculated as in Equation (17.6).

$$f = 100(v_S - v_R)/v_S \tag{17.5}$$

$$r = 100(1 - h_{out}/h_{in}) \tag{17.6}$$

where h_{in} and h_{out} are the entry and exit strip thickness, respectively.

The only dependence that showed some consistency was slip, but that dependence is built into Equation (17.7).

17.2 First-Principle Models

Three types of such models were reported for hot rolling. First, roll gap models were simplified to develop equations for the COF based on force and forward slip. In the second approach, the modelling was based on frictional stress rather than the COF. Finally, a team from the University of Sheffield developed a sophisticated stochastic model.

17.2.1 Modelling Involving Force and Forward Slip

Geffraye et al. (2000) and Sanfilippo et al. (2002) argued that it is better to base a model on slip than force, inasmuch as it is not sensitive to temperature, strip hardness and the velocity, diameter and Young modulus of rolls. A useful formula along these lines was devised by Carlton, Edwards, and Thomas (1976), and reported in its current form by McIntosh and Gunn (1992):

$$\mu = 0.54\sqrt{\frac{h_{in} - h_{out}}{R_{def}}} \Bigg/ \left[1 - \frac{1}{0.475}\sqrt{\frac{fh_{out}}{h_{in} - h_{out}}}\right] \tag{17.7}$$

where R_{def} is the deformed roll radius.

17.2.2 Modelling Based on Frictional Stress and Friction Factor

The exposition here follows Schey (1983). The coefficient of friction can be defined as:

$$\mu = F/P = \tau_i/p \tag{17.8}$$

where F is the force required to move the body, P normal force, τ_i the average frictional shear stress and p the normal pressure. This formulation is limited by sticking friction: when the frictional stress reaches the shear flow stress k, relative sliding at the interface stops. The workpiece is not glued to the tool, but as Schey succinctly describes it, shearing inside the workpiece requires less energy than for the workpiece to slide against the tool. The frictional shear stress can be then expressed as:

$$\tau_i = mk \tag{17.9}$$

where m is the frictional shear factor (0 at a frictionless interface, and 1 for sticking friction). Schey points out that k is usually known, whereas the normal pressure p has to be found, and depends on μ. However, it is difficult to assess m accurately from the bulk properties of the material. The smooth transition between the COF and friction factor models can be achieved by Wanheim–Bay or Stephenson friction models (Luo, 1995). Regarding the choice between the COF and frictional shear stress, Le and Sutcliffe (2002) believed that:

1. For relatively poor lubrication, where the frictional stress is due to the shearing of strip material at regions of strip/roll adhesion, the friction factor might be more suitable.
2. When lubrication conditions are relatively benign, the likely mechanism of friction is due to the strength of an interface layer (metal, oxide, soaps, lubricant) rather than metal strength. Shear strength of lubricants, and hence friction, tends to increase with pressure. So, COF could be more appropriate.

They also outlined another approach, which assumes that the average frictional stress τ is the sum of a stress τ_a due to boundary friction at asperity contacts, and a stress τ_v caused by the hydrodynamic shearing of lubricant:

$$\tau = A\tau_a + (1 - A)\tau_v \tag{17.10}$$

where A is the portion of contact surface where boundary friction dominates. The boundary friction stress is calculated using Coulomb's law, $\tau_a = \mu_a p_a$. The friction factor concept can be used to model the impact of contaminant films at the interface (Schey, 1983). The solid contaminant layer prohibits the welding of asperities. If its shear strength is τ_c, then $\tau_c = m_c k$. Assuming that factor m_c is known, which is rarely the case, as well as hardness H, the COF can be calculated as:

$$\mu = km_c \Big/ \left(H\sqrt{1 - m_c^2}\right) \tag{17.11}$$

17.2.3 Sheffield Model

Das, Palmiere, and Howard (2004) assumed that the interface in the roll gap between oxide, strip and roll can be represented as in Figure 17.1. The entities A–G are distinct probabilistic states, which are then combined in three groups, which makes the calculation of probability simpler:

1. AÉ: Strip is fully or partially in direct contact with roll, the probability of which is $P_{AÉ}$.
2. CEG: Strip is fully or partially in contact with roll or strip oxide (P_{CEG}).
3. BDF: Strip is in contact with air or lubricant (P_{BDF}).

Using metallographic data, the probabilities were estimated and expressed with the probability distribution diagram (PDD, Figure 17.2). If the roll gap pressure is below shear strength, the COF is:

$$\mu_{PDD} = P_{CEG}\mu_{oxide-metal} + P_{AE'}\mu_{metal-metal} \tag{17.12}$$

where the COF between oxide and metal, and metal and metal is assumed to be 0.1–0.2 and 0.3–0.4, respectively. At high pressure, 3–4 times the shear strength, the critical shear stress is:

$$\tau_{mod-crit} = P_{CEG}\tau_{oxide} + P_{AE'}\tau_{metal} \tag{17.13}$$

The results in Figure 17.3 suggest that the COF decreases with scale thickness, and increases after oxide changes from ductile to brittle. The model was not validated directly. Instead, a 3-D FE model simulated 'cut-grove'

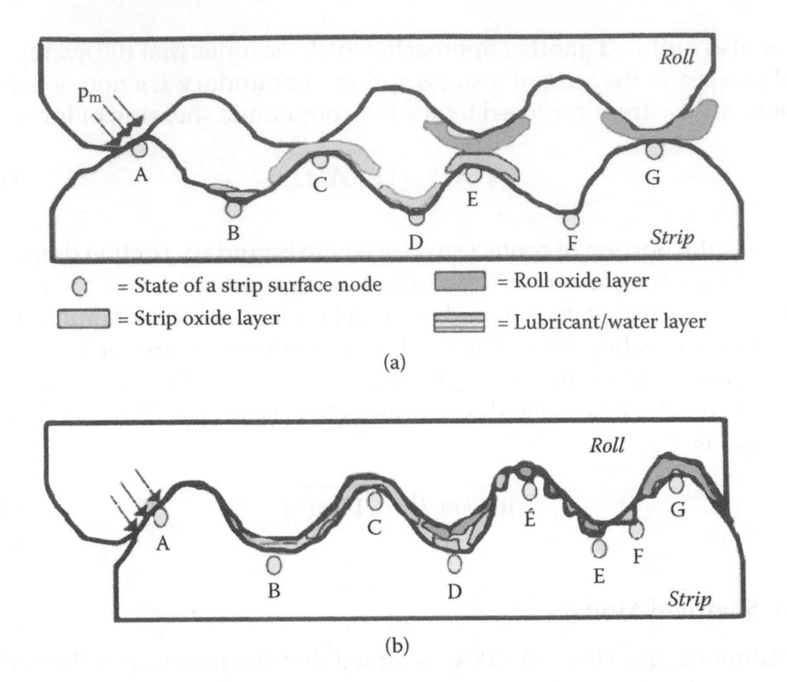

FIGURE 17.1
A schematic of two cases of the roll-strip-oxide interface, (a) the initial one, and (b) within the roll byte. The individual states are: (A) strip surface is in direct contact with roll and under contact pressure P_m; (B) strip surface is in contact with lubricant or water; (C) strip surface is in contact with strip oxide and under pressure; (D) strip surface is in contact with strip oxide; (E) strip oxide is in contact with pressurised roll oxide; (É) state developed due to the brittleness of the oxide layer below a critical ductile-brittle transition temperature; (F) strip is in contact with air; and (G) strip surface is in contact with pressurised roll oxide layer. (Reprinted from S. Das, E.J. Palmiere, and I.C. Howard, *Met. Mat. Trans.* 35A:1087–1095, 2004. With permission from Springer Verlag.)

experiments, and calculated 'grove opening' assuming either constant COF, or the one calculated with the PDD model. Better agreement between the calculated and measured grove opening was obtained with the PDD model.

17.3 Model Based on Commercial Mill Data

Given the evolution of COF over schedule as shown in Section 15.1, it is tempting to model the COF using the oxidation and wear of rolls. Hybrid modeling, combining empirically derived formulae with first-principle modelling, was conducted for stands F1–F4 in the commercial mill, because, as mentioned before, exit speed is not measured at stand F5:

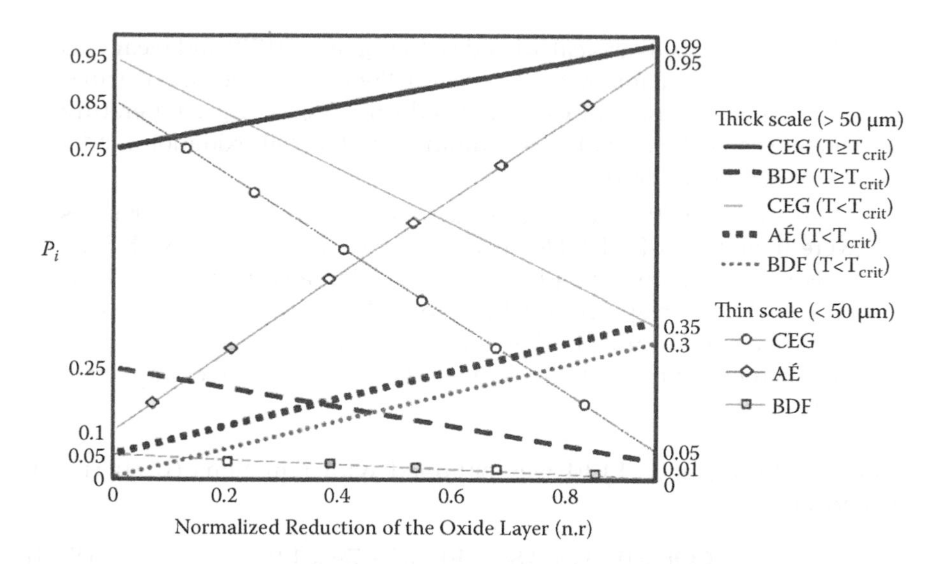

FIGURE 17.2
Probability distribution diagram with the probabilities of various groups. T_{crit} denotes the ductile-brittle transition temperature for oxide. (Reprinted from S. Das, E.J. Palmiere, and I.C. Howard, *Met. Mat. Trans.* 35A:1087–1095, 2004. With permission from Springer Verlag.) More details are given by S. Das, I.C. Howard, and E.J. Palmiere. *ISIJ Int.* 26:560–566 (2006).

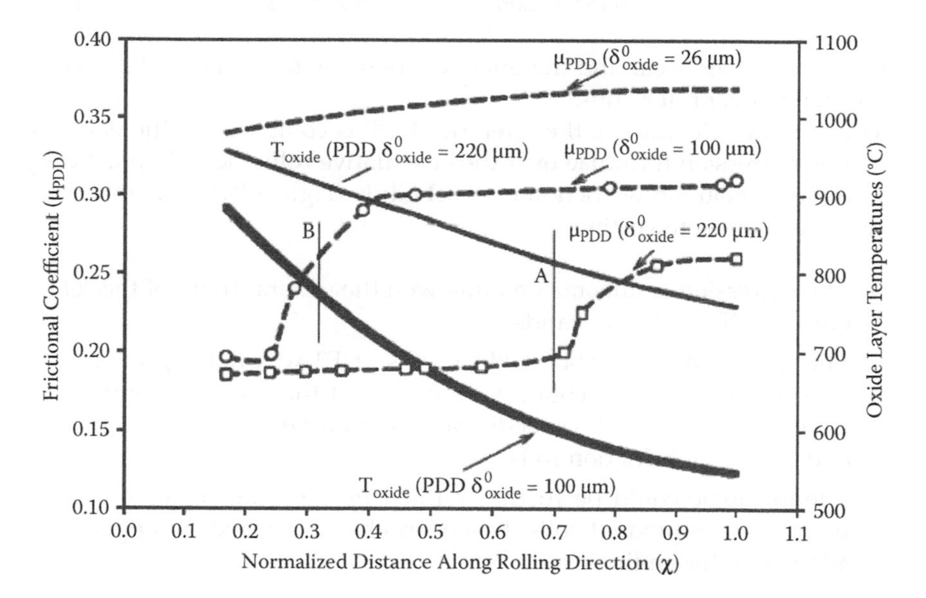

FIGURE 17.3
Frictional coefficient for different values of oxide thickness. A 220-μm-thick oxide film is brittle to the right of line A, and ductile to the left of it. (Reprinted from S. Das, E.J. Palmiere, and I.C. Howard, *Met. Mat. Trans.* 35A:1087–1095, 2004. With permission from Springer Verlag.)

1) The reference COF was calculated with Equation (17.7) and wear was estimated as in Panjkovic, Fraser, and Yuen (2004). For the calculation of the thickness of oxide formed on rolls, a strip temperature model (Panjkovic, 2007) was combined with a roll oxidation model (González et al., 2001).
2) More than 25,000 coils were used. Coils of all grades and gauges were included at F1–F3. However, high-strength–low-alloy (HSLA) coils were excluded at F4 because of the large scatter in the reference COF. The majority of HSLA coils did not produce these outliers, hence it is unlikely that the coil chemistry was a cause. It is more likely that the scatter was related to the larger forces applied to HSLA coils.

The best fit at stands F1-F4 was obtained with Equations (17.14)–(17.17), respectively:

$$\text{COF} = 0.333 + 3.88 \times 10^{-5}\, w^2 - 7.9 \times 10^{-4}\, \Omega \tag{17.14}$$

$$\text{COF} = 0.333 + 1.48 \times 10^{-5}\, w^2 - 9.6 \times 10^{-4}\, \Omega \tag{17.15}$$

$$\text{COF} = 0.219 + 1.3 \times 10^{-3}\, w - 4.4 \times 10^{-6}\, \Omega^2 \tag{17.16}$$

$$\text{COF} = 0.185 + 2.06 \times 10^{-4}\, w - 3 \times 10^{-4}\, \Omega \tag{17.17}$$

where w and Ω are wear and the thickness of oxide formed on rolls, respectively, in μm, over roll radius.

As a succinct illustration, the reference COF is compared to the one produced by regression formulae over the cumulative rolled length, practically showing the evolution of friction over schedule (Figure 17.4). The following observations are noteworthy:

1. The regression results match quite well the general trend of the reference COF at all four stands.
2. The 'kink', characteristic of HiCr rolls at F3 was not reproduced. Possibly the accuracy could be improved if the different formulae were developed for the specific ranges of cumulative rolled length and for the high-friction rolls.
3. The formulae could be further refined by deriving separate coefficients for individual rolls, based on chemistry, and by applying adaptation for on-line use.

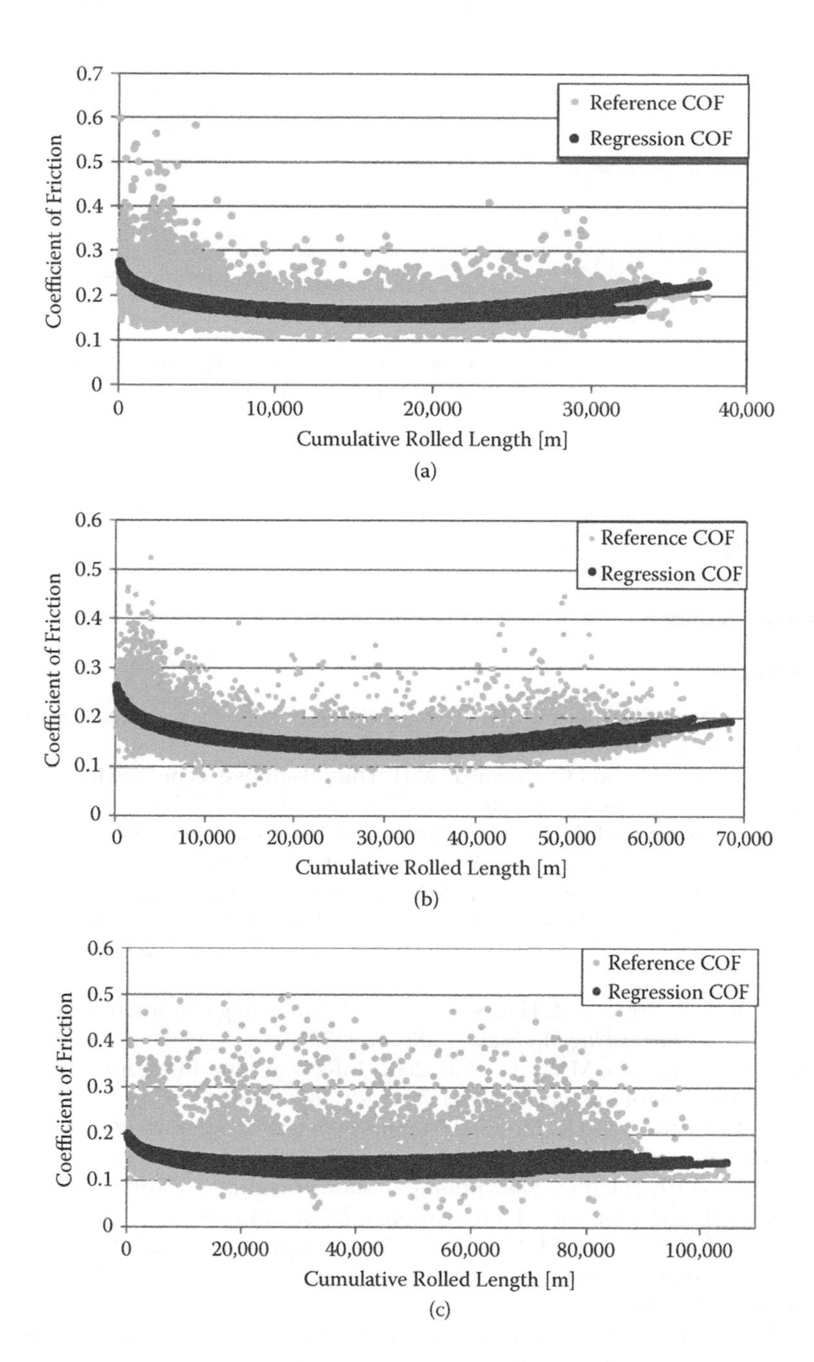

FIGURE 17.4
Comparison between the reference COF and the regression-calculated COF over the cumulative rolled length. (a) F1; (b) F2; (c) F3; and (d) F4.

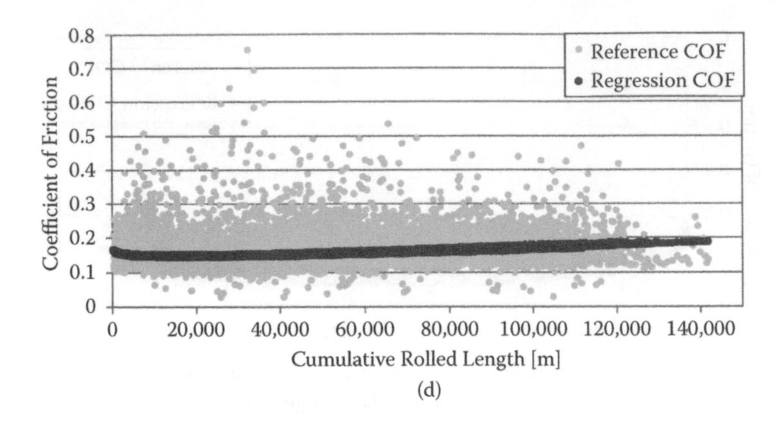

FIGURE 17.4 (continued)
Comparison between the reference COF and the regression-calculated COF over the cumulative rolled length. (a) F1; (b) F2; (c) F3; and (d) F4.

References

Carlton, A.J., W.J. Edwards, and P.J. Thomas. 1976. *Formulae for Cold Rolling Analysis.* John Lysaght: internal report.

Das, S., E.J. Palmiere, and I.C. Howard. 2004. The cut-groove technique to infer interfacial effects during hot rolling. *Met. Mat. Trans.* 35A:1087–1095.

Das, S., I.C. Howard, and E.J. Palmiere. 2006. A probabilistic approach to model interfacial phenomena during hot flat rolling of steels. *ISIJ Int.* 26:560–566.

Geffraye, F., V. Lanteri, P. Gratacos et al. 2000. Influence of the roll bite conditions on the surface quality of hot rolled coil. In *Proc. 42nd MWSP Conf.*, 233–242. ISS.

Ginzburg, V., and R. Ballas. 2000. *Fundamentals of Flat Rolling.* New York: Marcel Dekker.

González, V., P. Rodriguez, Z. Haduck et al. 2001. Modelling oxidation of hot rolling work rolls. *Ironmaking Steelmaking* 28:470–473.

Inoue, T., H. Yamamoto, M. Ataka et al. 2002. Effects of solid lubricants mixed with grease on hot rolling lubrication. *J. Jap. Soci. Tech. Plast.* 43:51–56.

Koseki, S., H. Yoshida, and K. Inoue. 1994. Improvement of accuracy of mathematical models for gauge set-up in hot strip finishing mills. *Tetsu-to-Hagane* 80:31–36.

Le, H.R., and M.P.F. Sutcliffe. 2002. Rolling of thin strip and foil: Application of a tribological model for 'mixed' lubrication. *J. Tribol.* 124:129–136.

Lee, W.H., J.H. Kwak, and C.J. Park. 1996. A new approach to predict rolling forces in the tandem cold rolling mill. In *Proc. 2nd Int. Conf. on Metal Rolling Processes*, 473–477. London.

Legrand, N., T. Lavalard, and A. Martins. 2012. New concept of friction sensor for strip rolling: Theoretical analysis. *Wear* 286-287:8–18.

Lenard, J.G., and L. Barbulovic-Nad. 2002. The coefficient of friction during hot rolling of low carbon steel strips. *Trans. ASME* 124:840–845.

Luo, C.-H. 1995. Modelling the frictional boundary conditions in a rolling process. *J. Mat. Proc. Tech.* 59: 373–380.

McIntosh, R.L., and B.M. Gunn. 1992. BHP Steel: internal report.

Milan, J.C.G., M.A. Carvalho, R.R. Xavier et al. 2005. Effect of temperature, normal load and pre-oxidation on the sliding wear of multi-component ferrous alloys. *Wear* 259:412–423.

Morales, J., I. Sandoval, and G. Murillo. 1999. Influence of process parameters on friction coefficient of high-chromium rolls. *AISE Steel Tech.* 76 (11):46–48.

Munther, P.A., and J.G. Lenard. 1997. A study of friction during hot rolling of steels. *Scand. J. Met.* 26:231–240.

Panjkovic, V. 2007. Model for prediction of strip temperature in hot strip steel mill. *App. Therm. Eng.* 27:2404–2414.

Panjkovic, V., G. Fraser, and D. Yuen. 2004. Applications of the crown and shape model in BlueScope Steel's Western Port hot strip mill. *Iron Steel Technol.* 1(10):98–107.

Sanfilippo, F., V. Lanteri, F. Geffraye et al. 2002. *Lubrication in Hot Rolling, Effect of Different Utilisation Strategies on Strip Quality and Process Conditions for Various Steel Grades*. Eur. Com., Rep. EUR 20208EN.

Sato, J., T. Kawashima, Y. Matsura et al. 1991. Advanced gauge control system for hot strip mill. *R&D Kobe Steel Eng. Rep.* 41(4):60–63.

Schey, J.A. 1983. *Tribology in Metalworking. Friction, Lubrication and Wear*. Ohio: American Society for Metals.

Vergne, C., D. Batazzi, C. Gaspard et al. 2006. Contribution of laboratory tribological investigations on the performance appraisal of work rolls for hot strip mill. In *Proc. ATS Rolling Conf., Paris, June 2006*.

Section IV

Appendices: Technical Details

Appendix A: Early Tribological Terminology in Great Britain

It seems that the term friction had not been used in England in its current meaning, as a physical phenomenon, until the late seventeenth century. Before then, and long afterwards, it simply denoted rubbing in medicine and natural sciences, and, according to Blau (2009, p. 2), it probably meant what is today known as frication, that is, 'the action of chafing or rubbing (the body) with the hands'. For example, the English translation of Francis Bacon's *Historia Vitae et Mortis*, published in 1669, contains: 'I have heard also of a Physician, yet living, who recovered a man to life which had hanged himself, and had hanged half an hour, by Frications and hot Baths ...' (p. 55).

He also believed that the gradual decline of humans occurs because: 'There is in every tangible body a spirit, covered and encompassed with the grosser parts of the body; and from it all Consumption and Dissolution taketh the beginning' (p. 57).

Now, to mitigate the bad influence of the spirit, he offered a remedy: 'The spirit equally dispersed maketh less haste to flie forth, and preyeth less upon the body, than unequally placed' (p. 60).

And this is best achieved in the following manner: '... therefore the more the spirit is shred and inserted by small portions, the less it preyeth; for Dissolution ever beginneth at that part where the spirit is looser. And therefore both Exercise and *Frications* conduce much to long life, for Agitation doth fineliest diffuse and commix things by small portions'.

Interestingly, the translation of the same work from 1858 uses the term *frictions*. Another example from the seventeenth century is Robert Boyle, who noted in 1663 that: 'Piso informs us, that the illiterate Brazilian empirics perform surprizing things both in the preservation of health, and the cure of many diseases, by their *frictions* in chronical, and unguents in acute cases' (Boyle, 1663a, p. 93).

Henry Oldenburg, the first secretary of the Royal Society (and the founding editor of its *Philosophical Transactions*, the longest running scientific journal) published in the first volume of the transactions, in 1665, an enthusiastic review of medical uses of friction: 'The operations and effects of touch and *friction* having been lately much taken notice off, and being lookt upon by some, as a great medical branch, for the curing of many diseases and infirmities ...' (p. 206).

Oldenburg quoted Boyle and Francis Bacon to support his views, and extolled the curative powers of friction by several examples, fairly amusing. We learn about 'a certain Cook in a noble family of England', who was troubled: '... having been reproached for the ugliness of his warty hands, and return'd for answer, that he had tried many remedies, but found none'. Finally, he '… was bid by his Lord, to rub his hand with that of a dead man' (p. 208).

In addition to the right advice, his Lord offered the right tool, too: '… and that his Lord dying soon after, the Cook made use both of his Lords advise and hand, and speedily found good effect' (p. 209).

This method performed beautifully on other occasions, and a certain Dr Harvey (presumably William Harvey, of the blood circulation fame) cured 'some tumors or excrescencies by holding on them such a hand'. The secret behind these miracles was revealed: 'Here is *friction* or touch, to mortifie wens, to drive away swellings and excrescencies' (Oldenburg, 1665, p. 208).

It was suggested that an even brighter future may be in store for friction: 'And why not to repell or dissipate spirits, that may have a dangerous influence upon the brain, or other parts; as well as to call forth the retired ones into the habit of the body, for invigoration?'

And there are other startling bits of evidence of the power of friction:

> … an aged Knight there [in Ireland], who having great pain in his feet, insomuch that he was unable to use them, suffered; as he was going to bed, a loving Spaniell to lick his feet; which was for the present very pleasing to him, so that he used it mornings and evenings, till he found the pain appeased, and the use of his feet restored.

> … he can assure of an honest Blacksmith, who by his healing hand converted his Bars of Iron into Plates of Silver; and had this particular faculty, that he caused Vomitings by stroaking the Stomack: gave the Stool by stroaking the Belly, appeased the Gout, and other paines by stroaking the parts affected.

The use of friction in a medical sense continued well into the nineteenth century. *Encyclopædia Britannica* from 1823 contains separate entries for the mechanical and medical meaning of friction, and the medical one is as follows: '*Friction*, in Medicine and Surgery, denotes the act of rubbing a diseased part with oils, unguents, or other matters, in order to ease, relieve, and cure it. *Frictions* are much used of late in venereal cases. They prefer the applying of mercury externally by way of *friction*, to that of giving it internally, to raise a salivation' (p. 229).

Whole books were published on the medical use of friction, with promising titles:

1. 'A full account of the system of *friction* as adopted and pursued with the greatest success in cases of contracted joints and lameness, from various causes' (Grosvenor and Cleoburey, 1825).
2. 'Every man his own doctor. The cold water, tepid water and *friction*-cure, as applicable to every disease to which the human frame is subject and also to the cure of disease in horses and cattle' (Claridge, 1849).

Regarding natural sciences, Boyle was using this term profusely in his essays on physics and chemistry. For example, he noted that: 'Sulphur, and amber are made, by *friction*, to omit odorable steams …' (Boyle, 1663a). He also observed that '… upon further trial with my own diamonds, by means of such a brisk and assiduous friction. I found that I could easily bring a diamond I wore on my finger to yield a sensible light …' (Boyle, 1663b). He reported that friction can 'made' diamond and glass electrical (Boyle, 1669), and in 1676 reported that steel tools can be magnetized by friction (Boyle, 1676).

In 1734, Desaguliers (1745) used 'stickage' as a synonym with friction, in line with his belief that adhesion is the cause of friction: '… we must make use of an additional Force, to overcome the *Stickage* arising from that Roughness. This *Friction* or *Stickage*, which is not great in other Engines, is very considerable in the Wedge. … we are to have regard to the imperfection of Engines and Materials, and the Quantity of *Stickage* or *Friction*' (pp. 93, 115).

This brief historical discourse is closed by defining lubrication. When mentioning lubrication, Hooke (1685) specifically mentions 'oil'. Rennie (1829) used the term *unguents* to describe lubricants generically, like Bennett in 1860 in his translation of Morin's famous book on mechanics. Jenkin and Ewing (1877) used this rather obscure term in the same sense. However, only nine years later, in his classical paper, Reynolds (1886) already used the modern term *lubrication*.

References

Bacon, F. 1669. *History Natural and Experimental of Life & Death: Or, of the Prolongation of Life*. London: William Lee. Originally published in Latin in 1623.

Bacon, F. 1858. *The works of Francis Bacon, Baron of Verulam, Viscount St. Alban, and Lord High Chancellor of England. Vol. V.* London: Longman.

Blau, P.J. 2009. *Friction Science and Technology*, 2nd ed. Boca Raton, FL: CRC Press.

Boyle, R. 1663a. The usefulness of experimental philosophy. In *The Philosophical Works of the Honourable Robert Boyle Esq.*, Vol. 1, 2nd ed., P. Shaw (Ed.), 1738, 1–173. London: W. Innys, R. Manby and T. Longman.

Boyle, R. 1663b. Promiscuous observations upon diamonds. In *The Philosophical Works of the Honourable Robert Boyle Esq.* Vol. 3, P. Shaw (Ed.), 1725, 144–147. London: W. and J. Innys, J. Osborn and T. Longman.

Boyle, R. 1669. The intestine motion of the parts of solids. In *The Philosophical Works of the Honourable Robert Boyle Esq.*, Vol. 1, 2nd ed. P. Shaw (Ed.), 1738, 160–163. London: W. Innys, R. Manby and T. Longman.

Boyle, R. 1676. The mechanical production of magnetism. In *The Philosophical Works of the Honourable Robert Boyle Esq.*, Vol. 1, 2nd ed. P. Shaw (Ed.), 1738, 496–505. London: W. Innys, R. Manby and T. Longman.

Claridge, R.T. 1849. *Every man his own doctor. The cold water, tepid water and friction-cure, as applicable to every disease to which the human frame is subject and also to the cure of disease in horses and cattle*. London: James Madden.

Desaguliers, J.T. 1745. *A Course of Experimental Philosophy, Vol. I*, 2nd ed. London: Innys, Longman, Shewell, Hitch and Senex.

Encyclopædia Britannica Vol. IX, 6th ed. 1823. Edinburgh: Constable.

Grosvenor, J., and W. Cleoburey. 1825. *A full account of the system of friction as adopted and pursued with the greatest success in cases of contracted joints and lameness, from various causes*. Oxford: Munday and Slatter.

Hooke, R. 1726. *Discourse of carriages before the Royal Society, on Feb. 25 1685. With a description of Stevin's Sailing Chariot, made for the Prince of Orange. In Philosophical Experiments and Observations of the Late Eminent Dr. Robert Hooke*, 1726, 150–167. London: W. Derham and J. Innys.

Jenkin, F., and J.A. Ewing. 1877. On friction between surfaces moving at low speeds. *Phil. Trans*. 167:509–528.

Morin, A. 1860. *Fundamental Ideas of Mechanics and Experimental Data*. Translated by J. Bennett. New York: D. Appleton.

Oldenburg, H. 1665. Some observations of the effects of touch and friction. *Phil. Trans*. 1:206–209.

Rennie, G. 1829. Experiments on the friction and abrasion of the surfaces of solids. *Phil. Trans*. 119:143–170.

Reynolds, O. 1886. On the theory of lubrication and its application to Mr. Beauchamp Tower's experiments, including an experimental determination of the viscosity of olive oil. *Phil. Trans*. 177:157–234.

Appendix B: Some Clarifications of the Stribeck Curve

1. *The name.* The curve is named after Stribeck, but Fusaro (1991) calls it Stribeck–Hersey, and Bowden and Tabor (1950) use it as a nameless graph. Sorooshian (2005) calls it Stribeck–Gümbel, as does Ludema (1996), who also calls it McKee–Petroff.
2. *The parameter on the abscissa.* Generally, it is a product of viscosity and sliding speed, divided by the load. However, there are other definitions, and several names are given to the number. Those better known are summarised in Table B.1.
3. *The symbol 'Z'.* This symbol (or *z*) often denotes viscosity, after the German word *Zählkeit* (Dowson, 1998). This term in modern German denotes resistance to deformation in general, of liquids and solids alike.

TABLE B.1
Parameters Used in the Stribeck Curve

Parameter (Number) on Abscissa	Name of the Number	Typical Use	Source
$\frac{\eta N}{p}\left(\frac{L}{c}\right)^2$	Sommerfeld	Journal bearings	Ludema, 1996
$\frac{P}{\eta U}\left(\frac{c}{r}\right)^2$	Sommerfeld	Journal bearings	Blau, 1992
$\frac{\eta N}{p}\left(\frac{r}{c}\right)^2$	Sommerfeld	Heavily loaded journal bearings	Blau, 1992
$hu\eta/p$	Sommerfeld		Sorooshian, 2005
$\frac{u\eta}{pb}$	Sommerfeld	Piston rings-liner contacts	Ludema, 1996
$u\eta/p$	Hersey		Blau, 1992 Sorooshian, 2005
ZN/p	Hersey		Ludema, 1996
$\frac{P}{\eta U}$	Hershey [sic!]		Blau, 1992
$v\eta/w$	Stribeck	Cylinder-plane contacts	Ludema, 1996
$\frac{\eta N}{p}\left(\frac{d}{b}\right)^2$	Ocvirk	Journal bearings	Blau, 1992
ZN/p	Unnamed		Bowden and Tabor, 1950

Note: ab, Ring width or bearing length c, bearing clearance; d, bearing diameter; h, film thickness; L, bearing length; N, frequency of rotation; p, pressure; P, load per unit width; r, bearing radius; u, sliding velocity; U, surface velocity; v, entraining velocity; w, specific force over the width of cylindrical specimen; Z, η, dynamic viscosity.

References

Blau, P.J. 1992. Glossary of terms. In *Friction, Lubrication, and Wear Technology, ASM Handbook, Vol. 18*, 1–21. Ohio American Society for Metals.

Bowden, F.P., and D. Tabor. 1950. *The Friction and Lubrication of Solids*. Oxford: Oxford University Press.

Dowson, D. 1998. *History of Tribology,* 2nd ed. London and Bury St Edmunds: Professional Engineering.

Fusaro, R.L. 1991. *Tribology Needs for Future Space and Aeronautical Systems*. NASA TM 104525.

Ludema, K. 1996. *Friction, Wear, Lubrication*. Boca Raton, FL: CRC Press.

Sorooshian, J. 2005. *Tribological, Thermal and Kinetic Characterization of Dielectric and Metal Chemical Mechanical Planarization Process*. PhD diss., University of Arizona.

Appendix C: Bowden and Tabor—Life and Work

Bowden (Figure C.1) and Tabor (Figure C.2) are fascinating in many ways. First, they made an enormous contribution to tribology. Second, they had an exemplary approach to the creation of links between practical problems and fundamental research. Third, Bowden was very successful in bringing industrial and military funding for fundamental research. Fourth, by good scientific work they made a significant technical contribution to the Australian war effort in WW II in the Lubricants and Bearings Section of CSIR (Council for Scientific and Industrial Research).

C.1 Frank Philip Bowden, the Greatest Australian Tribologist*

> Bowden was a man of many talents and could have made his mark as a writer, as a lecturer, as an aesthete, as a politician and statesman, as an administrator and man of affairs, as a scientist ...
>
> **Tabor, 1969**

Before Yahoo Serious played Young Einstein in the eponymous movie, and Frances Blackburn was awarded the Nobel Prize in medicine, the best known Tasmanian-born scientist had been Frank Philip Bowden (1903–1968). Bowden did not leave any autobiographical notes, and this biography is mainly based on the obituary prepared by Tabor (1969), unless indicated otherwise.

Born in 1903 in Hobart, Bowden had an inauspicious start. Although a good pupil, he was not brilliant and failed in mathematics in high school, so he could not enter a university†. Subsequently, he found employment as a junior laboratory assistant in the Electrolytic Zinc Company, and started developing a taste for experimental work. His abilities were quickly noticed, and the company staff persuaded him to continue with education. He eventually matriculated and enrolled at the University of Tasmania in 1921 to

* Some may argue that Anthony George Maldon Michell (Cherry, 1962; Dowson, 1998), of the tilting-pad thrust-bearings fame, deserves the title. Maldon was a world-class tribologist, 'arguably Australia's most versatile engineer' (Walker, 1986), and a Royal Society Fellow.

† Bowden was never good at mathematics, hence the absence of sophisticated mathematical models in his papers.

FIGURE C.1
Frank Philip Bowden. (Reprinted with permission from Lafayette Photography, Cambridge.)

study science. Again, things were not going smoothly. He fell ill and was advised to spend six months in a warmer climate. He was sent to a station somewhere in the country of New South Wales, where he rode horses, hunted kangaroos and worked as a jackaroo.

Once he recuperated, the progress in studies was swift; a BSc degree was obtained in 1924, and a MSc degree (with first honours) in 1925. An interesting anecdote from this period of his life is noteworthy. His physics teacher, Alexander Leicester McAulay*, wanted to obtain a grant for Bowden's research from the Electrolytic Zinc Company. He found it hard to meet the company manager and negotiate the grant. Finally, he heard that every morning that manager was having a shave in a particular barbershop. McAulay cornered the manager there and secured the coveted funding, in negotiations dominated by '… the sound of the razor scraping away the managerial beard'.

In 1927, Bowden left for Cambridge to study for his PhD. He worked mainly in electrochemistry, eventually moving towards the studies of friction. There he was quickly noted for his ability to devise 'conceptually simple experiments that went to the heart of the problem' (Greenwood and Spink, 2003). Soon he demonstrated his knack for collaboration with industry and

* According to Scott (1986), McAulay's exterior was unusual: '… tall and thin with hawk-like features, long untidy hair and shabby clothes', and would occasionally use a piece of string instead of a belt on his trousers. However, McAulay was a practical man, and, like Bowden, contributed much to the Australian war effort. There was a critical shortage of optical components in Australia for military use in 1940. Without having any experts in his department '… Leicester built up a team which short-cut procedures and within months was producing precision prisms and lenses for gun-sights and cameras'.

the military. In his mid-thirties, he was consulting for Shell, and under his supervision the company established a small research outfit for wear and lubrication. Then he collaborated with the British Air Ministry, the Fuel Research Board and the War Office.

The outbreak of the Second World War found Bowden in Australia, while he was returning to the United Kingdom from a lecturing tour in the United States. He decided to establish a small group to offer technical support to the war effort. Both the formation of the group and its achievement were remarkable (see Section C.3). In 1945 Bowden resigned from CSIR and returned to Cambridge, where he established a research group mainly using grants from the British Ministry of Supply. The group worked on friction, lubrication and the initiation and detonation of explosives, and later expanded into other fields, to name a few: the damage inflicted on solids by supersonic rain drops, the role of stress-wave in the deformation of brittle solids, high strain-rate phenomena, the range of action of surface forces and so on.

Back in Cambridge, Bowden strengthened links with industry. He advised in 1954 on the set-up and running of research facilities in Tube Investments Ltd (Melford et al., 2010). In 1958, English Electric Company elected him a director. Eventually he became the chairman of the company's research council. It is not specified what made him that successful, though details emerge in Sections C.3 and C.4, as glimpsed from Tabor (1969) and Greenwood and Spink (2003). Obviously, he had enough high-class charm to mingle with important business and military persons, and knew how to handle them. For example, his office in Cavendish Laboratory was 'magnificent' and 'capacious' (Field, 2008). Another, much better known luminary of Cavendish, Lord Rutherford, had a surprisingly small and cramped office (Cathcart, 2005). This does not mean that Bowden had a particular penchant for grandeur; an impressive office was certainly an asset when dealing with senior figures in a class-conscious society. He was a successful team leader, and paid close attention to the personal and professional well-being of the staff. After careful consideration of a problem, often with a humourous comment, Bowden would effectively solve a complicated research or personnel problem. Also, he was an astute judge of men and affairs, well read, with a wide range of cultural interests. Finally, what helped was his keen interest in industrial problems, an ability to identify fundamental scientific aspects of practical problems, and an uncanny talent to apply fundamental research to real-life technical problems.

By the mid-1950s he was recognised world-wide as a leading tribologist, which is best illustrated by the following detail. In 1956 Kragelsky and Shchedrov published in Russian the first book dedicated to the history of tribology. In the author index of the book, only Coulomb (47) and Amontons (31) scored more mentions than Bowden (24).

Bowden's death came rather early; he died of cancer in 1968. He left behind a legacy of more than 170 journal papers, several monographs and the timeless book *The Friction and Lubrication of Solids*, co-authored with Tabor. The range of topics covered in his papers is overwhelming, from electrochemistry and skiing, over tribology and explosives, to orthopaedics and experimental biology (e.g., 'Effect of irradiation with different wave-lengths on the oestrous cycle of the ferret with remarks on the factors controlling sexual periodicity', published in 1934).

C.2 David Tabor, a Quiet Achiever

According to the obituary compiled by Field (2008), Tabor (1913–2005, born Tabrisky) had an interesting background. His father Ezekiel was a metalworker in Russia, who obtained employment in the army. When the Czar's uncle visited his unit, he prepared an exhibition of arms in the shape of the Russian Eagle. The Prince was mightily impressed and immediately required to talk to him. After noticing that Ezekiel was a Jew, and therefore not permitted to serve in the Russian Army, the Prince demanded that Ezekiel convert to Christianity. The latter was not overly enthusiastic about it, so he had to quit the army and start his own business. Fortunately, his commanding officer used his services afterwards, even going to the trouble of supplying Ezekiel with the papers necessary to migrate to Great Britain. There Ezekiel set up a metalworking business, and anglicised his name to Charles Tabor, sometime after the birth of David in London.

FIGURE C.2
David Tabor. (Reprinted from J. Field, *Biogr. Mems. Fell. R. Soc.* 54:425–459, 2008. With permission from the Royal Society.)

Like Bowden, he was struck with a disease in his youth. He survived osteomyelitis, though with a leg slightly shorter than the other. He had to rely on a surgical boot and walking stick, but enjoyed hiking, swimming and even tennis. He was a good student and won a scholarship to Imperial College, graduating in physics. He stayed there, and once failed to replicate some results obtained by William Hardy (of boundary lubrication fame). He was advised to visit Bowden, who had encountered a similar problem. Bowden already had formed a research group, and offered Tabor to join the group and study for his PhD. This was the start of an effective collaboration that lasted more than 30 years.

Bowden formed his wartime research group in Melbourne in late 1939, and Tabor joined in early 1940, thinking that the war would end soon. He stayed in Melbourne until 1946 and renamed the group 'Division of tribophysics'. It is quite possible that this name inspired the term 'tribology', adopted in 1966 in the so-called Jost report. According to Dowson (1998), Bowden was consulted about the choice of name.

Back in Cambridge, Tabor spent half a century in productive research work in tribology and surface physics in general, publishing papers until 1998. Judging by the number of contributors to his obituary, he had many students who respected and liked him. By all accounts, he was a very competent scientist, kind and patient with students, and treating them as 'intellectual equals'. Unlike Bowden, he possessed extraordinary writing skills and was a good lecturer in academia. Above all, he was exceptionally modest, and rarely put his name first on joint publications, even when he made the largest contribution. He did not add his name to the Johnson–Kendall–Roberts theory of adhesion, developed by his eponymous students, despite his significant input. He was always tactful and considerate in dealing with peers and students. As Freitag recollects in the obituary, critique was given 'not as a rebuke but rather as a nudge with a hint of how a better result may be obtained'. Problematic issues of someone's work were not addressed in front of an audience; instead, they were 'discussed in a smaller circle on a suitable occasion'.

Interestingly, Tabor lived a very long life despite being a smoker. In the obituary, principal assistant Arthur Stripe remembers Tabor's numerous comic attempts to quit smoking. He would stop bringing cigarettes to work, but would stop the first smoker he came across and ask for a cigarette, saying, 'I am giving up, but not quite yet'.

C.3 Lubricants and Bearings Section of CSIR

Bowden's successful effort to found a laboratory in Melbourne and carry out research of interest to the Australian war effort illustrates his outstanding managerial skills. While returning from a lecturing trip in United States to England, he was in Australia when the war broke out in September 1939. Bowden decided to support his country by setting up a small outfit to offer technical support to various aspects of the war industry and military technology. The following story is based on the reminiscences collected by Tabor (1969), Greenwood and Spink (2003) and Field (2008).

He quickly organised discussions with Aeronautical Research Laboratories, Holden, Army's Mechanical Transport, Department of Supply, Commonwealth Aircraft Corporation, RAAF and the University of Melbourne. He also presented his Cambridge work to many scientists, engineers and industrialists. The presentation was well received and many supported the formation of a research group. Bowden wrote a memorandum to the head of CSIR (a predecessor of CSIRO), Sir David Rivett. According to Tabor (1969), the memorandum was 'a typical Bowdenesque document', forthright, succinct, convincing, with 'a modest but firm statement of requirements'.

Rivett passed the proposal to the Minister in Charge of CSIR, Richard Casey, and then the going became tough. Casey was the Minister for Supply and Development in Menzies' government, and an engineer by trade. He found the proposal interesting, but thought that Bowden better return to his fundamental work in Cambridge. To add insult to injury, an unnamed industrialist claimed Bowden to be 'too airy-fairy'. Casey (Hudson, 1993) was a formidable obstacle, being a distinguished man (politician, diplomat, the sixteenth governor-general of Australia, and so on*). Worse, Casey referred the case to the renowned businessman Sir Colin Fraser, chairman of the Advisory Panel on Industrial Organization (a body planning the conversion of Australian private industry to war-time production), who strongly opposed the idea†. The panel included some crusty industrialists, and this opposition is hardly surprising.

However, Bowden was undaunted, and was assisted by Rivett in pressing Casey, who eventually invited Bowden for an interview. Bowden was eager and prepared, and presented his case well (with 'a forceful combination of charm, persistence and a manifest commitment to solving practical problems

* Casey was the right-hand man of Sir Robert Menzies, and played very important roles in the creation of the South-East Asia Collective Defence Treaty in 1955 and the Australian–Japanese Trade Agreements in the 1950s (Menzies, 1970).

† Greenwood and Spink state that Fraser was, at that time, the chairman of the panel. So, the events definitely took the place right after the handover of the chairmanship from the legendary boss of Broken Hill Proprietary Ltd, Essington Lewis, who was still the panel chairman in September 1939 (Blainey, 1971). Lewis certainly had the clout on the panel, and Fraser and Lewis were close. When Lewis became the Director-General of Munitions, he appointed Fraser the Director of Materials Supply. Lewis' views on research were negative (Blainey, 1971): 'He saw little merit in employing men solely in the hope that, after many years' work, they might alight on a discovery. "Let someone else pay for the mistakes", he once told … his technical assistant. Lewis resented mistakes, and in research many mistakes were unavoidable.'

of immediate concern to war effort' (Greenwood and Spink, 2003). Casey was convinced that the proposal had a definite practical value*. Bowden became Officer-in-Charge of Lubricants and Bearings Section of CSIR a month later†. This section was located on the grounds of the University of Melbourne, and cooperated well with the academics. This further underlines Bowden's management qualities. In those days, the relationship between academia and CSIR was less than harmonious. For, 'University departments were jealous of the comparatively lavish equipment, technical assistance, and buildings provided for CSIR workers, while they lacked the most elementary resources for their research' (Marston, 1966).

His team worked in four key areas (1) friction, lubrication, bearings and wear; (2) initiation and propagation of explosions; (3) muzzle velocity of projectiles; and (4) physical metallurgy (Greenwood and Spink, 2003). The section grew up to almost a couple of dozens of people by the end of the war, and produced an outstanding record of achievements. It suffices here to say that the production of aircraft in Australia critically depended on the technical expertise of Bowden's team, particularly in the area of the casting of bearings. This enterprise alone would suffice as the ultimate proof of Bowden's management abilities. But as seen in Section C.2, he demonstrated his leadership skills many times over.

C.4 Bowden and Tabor, or Tabor and Bowden?

Bowden and Tabor formed an inseparable team, and are very often cited together, with Bowden commonly mentioned first. That makes one assume that their scientific contribution was about equal, with Bowden's slightly larger. However, more recent articles indicate that rather than giving equal scientific output, their contributions were different in nature and size, but perfectly complemented each other.

Although both Bowden and Tabor are often credited for the modern adhesion theory of friction, some authors assign it solely to Tabor. When praising Merchant's adhesion theory of friction, Bisson (1968) and Komanduri (2006) contend that it was developed independently of and simultaneously with Tabor, without mentioning Bowden. While defending the priority of Bowden and Tabor, Ludema (1996) remarked that '... Bowden and Tabor are worthy of the honour accorded them. Bowden for his prowess in acquiring funds for the laboratory and Tabor for the actual development of concepts'.

* Interestingly, Casey was well acquainted with the other great Australian tribologist, Michell (Cherry, 1962); perhaps that also helped Bowden.

† Bowden deliberately chose a plain, unassuming name to underline the practical aspect of work and avoid projecting a boffin image.

In Tabor's obituary (Field, 2008), Freitag recollects that Tabor had the role of 'scientific conscience and scientific communicator', whereas Bowden 'provided the resources and the connections to industry and politics, and also frequently injected ideas for future projects'. He reveals that Tabor practically wrote the first volume of the famous book by Bowden and Tabor, *The Friction and Lubrication of Solids,* save the preface. Tabor often put his students as the first authors of joint papers, even when he mainly wrote them. On the other hand, Bowden was listed as the first author of the vast majority of joint publications with others, even when it was obvious that, due to his many duties, he could not be the dominant contributor.

However, it would be wrong to conclude that Bowden was freeloading at Tabor's expense. In the same obituary, Hutchings notes that their relationship was harmonious, because they respected each other's qualities. Tabor was happy to leave the management and funding chores to Bowden, and to focus on research and writing. He credits Bowden for running not only a successful, but 'happy' research department. The absence of financial worries helped the staff to focus on work, and they must have been grateful to Bowden for that.

It would be also wrong to conclude that Bowden was a scientific lightweight. As a young scientist, he abundantly demonstrated his scientific abilities. Later, as a scientific leader, Bowden generated ideas for the new areas of research, and saw the big picture of friction research. Tabor himself credited him with the rare talent not only to 'find and "skim the cream off" interesting problems', but also to 'find cream where nobody else could'.

Maybe we can say Tabor and Bowden, rather than Bowden and Tabor. However, my view is that Bowden was to tribology what Robert Oppenheimer (Bethe, 1968) and John Cockcroft (Oliphant and Penney, 1968; Conant, 2003) were in the US nuclear program, and British nuclear physics and the wartime radar development, respectively. They were outstanding scientific experts in their field, and, at the same time, great administrators, capable of successfully running large-scale projects, earning the respect of both their subordinates and superiors. Maybe all three are best described by the attributes assigned to Cockcroft at his memorial service by Spence (1967): '... scientist, creator and administrator of great projects, and technological statesman'.

C.5 Exemplary Approach to Industrial Research by Bowden and His Teams

Both Bowden and Tabor had 'a flair to see the practical advantages of their research to industry' (Field, 2008). Above that, Bowden was an outstanding expert in earning the trust and funding from industry, military and government. Israelachvili recollects that as a prospective PhD student he 'was told to visit Professor Bowden's department because he had lots of unrestricted money from industry' (Field, 2008). At the same time, Bowden, Tabor and collaborators achieved an enviable publishing record, with dozens of papers published not only in the journals of the Royal Society, but many more elsewhere. It was a uniquely successful combination of fundamental and applied research. It appears that the outstanding success rested on several pillars, listed below in no particular order of merit:

1. *Rigourous and methodical scientific approach to practical problems.* The group was involved in the investigation of the explosion in the ammunitions factory in Deer Park (Greenwood and Spink, 2003), and the details were supplied by Tabor (1969). One approach involved dropping steel balls from a certain height on the droplets of nitroglycerine on an anvil. However, explosions were rare. Someone suggested that the nitroglycerine escaped during impact. The cavities were placed into the nose of impacting solids, which increased the sensitivity to impact, but the results were still erratic. Bowden then suggested that tiny gas bubbles should be introduced in the explosive. Finally, the sensitivity became high and the explosions were reproducible. This suggested that the key cause was the generation of hot spots by the adiabatic compression of trapped gas bubbles. Two additional causes were established later, namely 'viscous heating during the extrusion of liquid films between heavily impacting surfaces', and the creation of hot spots via frictional heating. It was shown that the hot spot temperature of 500°C could trigger an explosion.
2. *Rapid transfer of solutions developed in one research field to another.* The possibility to increase the penetration of bullets through metal sheets by lubricating the bullet nose was considered. This required the determination of bullet velocity, and a team member (Jeofry Courtney-Pratt) devised a simple and effective apparatus to measure it. This attracted the attention of the army personnel, and the equipment was modified and applied to the calibration of the large guns on the ships of the Royal Australian Navy (Tabor, 1969).
3. *Bold deployment of research results to industry.* An example is the work of Bowden and Tabor on the use of films of soft metals to reduce friction (Field, 2009). For example, use of lead on brass could reduce friction by an order of magnitude. Tabor once asked an operator in an ammunitions factory to remove all the lubricant from the die, and to pass through it a brass shell coated with lead. The operator was alarmed, fearing that the die would be damaged. 'Tabor told him he had authority to do so. With incredulity, the operator put the piece in the die and activated the plunger. It went through like a dream' (Field, 2008).

4. *Close links with industry*. Tabor (1969) noted that Bowden had very strong feelings about establishments where researchers had weak links with the development and manufacturing departments. He was not that much concerned with the 'ivory tower' mentality of scientists, as with the resentment that could be generated among production personnel that 'research people were being accorded privileged treatment', without having to contribute much to pressing problems.
5. *Awareness of the mutual dependence of industry and research*. According to Tabor (1969), Bowden relished being, even temporarily, in charge of research facilities in Tube Investments Ltd. He firmly believed that the increase of the pool of scientific knowledge and the commercial benefits of the company funding a research facility fed each other. He also encouraged the focussing of research work on the issues that could bring the greatest financial benefits.
6. *Recruitment of outstanding performers*. Bowden was capable of recruiting people who later went on to internationally recognised careers. The careers of team members in the Lubricants and Bearings Section are a good example, and several examples are provided by Greenwood and Spinks (2003).

References

Bethe, H.A. 1968. J. Robert Oppenheimer. 1904–1967. *Biogr. Mems Fell. R. Soc*. 14:390–416.

Bisson, E.E. 1968. *Friction, Wear and the Influence of Surfaces*. NASA TM/X-52380.

Blainey, G. 1971. *The Steel Master: A Life of Essington Lewis*. Melbourne: Macmillan of Australia.

Cathcart, B. 2005. *The Fly in the Cathedral*. New York: Farrar, Straus and Giroux.

Cherry, T.M. 1962. Anthony George Maldon Michell. 1870–1959. *Biogr. Mems. Fell. R. Soc*. 8:90–103.

Conant, J. 2003. *Tuxedo Park: A Wall Street Tycoon and the Secret Palace of Science That Changed the Course of World War II*. New York: Simon and Schuster.

Dowson, D. 1998. *History of Tribology*, 2nd ed. London and Bury St Edmunds: Professional Engineering.

Field, J. 2008. David Tabor. 23 October 1913–26 November 2005. *Biogr. Mems. Fell. R. Soc*. 54:425–459.

Greenwood, N.N., and J.A. Spink. 2003. An antipodean laboratory of remarkable distinction. *Notes Rec. R. Soc*. 57:85–105.

Hudson, W. 1993. Casey, Richard Gavin Gardiner (Baron Casey) (1890–1976). In *Australian Dictionary of Biography, Vol. 13*, 381–385. Melbourne: Melbourne University Press.

Komanduri, R. 2006. In Memoriam: M. Eugene Merchant. *Trans. ASME* 128:1034–1036.

Kragelsky, I.V., and V.S. Shchedrov. 1956. *Development of the Science of Friction: Dry Friction*. Moscow: Academy of Sciences of USSR.
Ludema, K. 1996. *Friction, Wear, Lubrication*. Boca Raton, FL: CRC Press.
Marston, H.R. 1969. Albert Cherbury David Rivett. 1885–1961. *Biogr. Mems Fell. R. Soc.* 15:437–455.
Melford, D., P. Duncumb, M. Stowell et al. 2010. Tube Investments Group research laboratory, Hinxton Hall (1954–88). *Notes Rec. R. Soc.* 64:287–310.
Menzies, R. 1970. *The Measure of the Years*. London: Cassell.
Oliphant, M.L.E., and W.G. Penney. 1968. John Douglas Cockcroft. 1897–1967. *Biogr. Mems Fell. R. Soc.* 14:139–188.
Scott, B. 1986. McAulay, Alexander Leicester (1895–1969). In *Australian Dictionary of Biography, Vol. 10*, 202–204. Melbourne: Melbourne University Press.
Spence, R. 1968. Address at the service of memorial and thanksgiving for Sir John Cockcroft, O.M., K.C.B., F.R.S. at Westminster Abbey on 17 October 1967. *Notes Rec. R. Soc.* 23:31–32.
Tabor, D. 1969. Frank Philip Bowden. 1903–1968. *Biogr. Mems Fell. R. Soc.* 15:1–38.
Tabor, D. 2005. Obituary. http://www.smf.phy.cam.ac.UK/files/574FrictFieldBMFRS54.pdf
Walker, S. 1986. Michell, Anthony George Maldon (1870–1959). In *Australian Dictionary of Biography, Vol. 10*, 492–494. Melbourne: Melbourne University Press.

Appendix D: Properties of Metals, Oxides, Carbides and Rolls

The adhesion theory of friction dictates that an efficient lubricating film must have lower shear strength than metallic substrates, and similar hardness. Application of this theory to rolling requires knowledge of these properties of oxides, carbides and rolls. These data are dispersed throughout the literature, and it is useful to collect them in one place.

Properties of metals and oxides relevant to work rolls are compiled in Tables D.1 and D.2. Lundberg and Gustafsson (1994) proposed formulae for the calculation of the hardness of iron oxides as a function of temperature, and the resulting curves are shown in Figure D.1. It shows that hæmatite is the hardest, and wüstite the softest of iron oxides, whereas iron is softer than any of its oxides, although experimental data by Vagnard and Manenc (1964) show that wüstite is softer than pure iron in the range of 950°C–1000°C.

Due to the lack of specific information in the sources, properties in the tables are presumed to be at room temperature, unless specified otherwise. It is hard to find data on strip and oxide hardness and strength at elevated temperatures, but the situation is somewhat better with roll materials. However, the differences between roll and oxide properties, and between strip and oxide strength at mill temperatures are large enough to allow a qualitative analysis.

Hardness of carbides is listed in Table D.3. Many authors quote the hardness of generic carbides, that is, those presented as M_xC_y. The range of these data is wide, because hardness of carbides depends on their metallic component. However, these data are still useful, and are compiled in Table D.4. An interesting comparison of the hardness of various carbides is shown in Figure D.2. Composition of those carbides found in rolls is given in Table D.5.

Various useful roll data are listed in Table D.6. The dependence of the hardness of roll matrix on temperature is illustrated in Figure D.3, and the curves shown agree well with the dependence of roll bulk hardness on temperature (Hashimoto et al., 1995; Lecomte-Beckers, Terziev, and Breyer, 1997). Generally, HSS rolls are harder than HiCr due to harder matrix and the presence of MC carbides, harder than M_7C_3 carbides in HiCr rolls (Pellizzari, Cescato, and De Flora, 2009). However, it can be seen in Table D.6 that the hardness of HSS and HiCr rolls is similar at 600°C, which is the upper limit of roll surface temperature in hot rolling.

TABLE D.1
Physical Properties of Metals and Oxides

Metals, Oxides	Density (kg m^{-3})	Hardness	Shear Strength (MPa)	Tensile Strength (MPa)[a]	Reference
C-steel			1140		Stott and Jordan (2001)
				160[b]	Anon. (1990)
Fe			980		Buckley (1971)
FeO		460 HV[c]			Luong and Heijkoop (1981)
		270–350 HV[c]			Picqué (2004)
Fe_3O_4			550		Eadie et al. (2002)
		540 HV[c]			Luong and Heijkoop (1981)
		420–450 HV[c]			Picqué (2004)
Fe_2O_3		1050 HV[c]			
		1030 HV[c]	500		
			1640		Lu et al. (2005)
					Buckley (1971)
					Luong and Heijkoop (1981)
					Picqué (2004)
Cr			1200		Buckley (1971)
Cr_2O_3	4900	900–1500 VPN			
		1800 kg mm^{-2}	1310		Mann and Prakash (2000)
					Sliney (1991)
					Buckley (1971)
Mo			1190		Buckley (1971)
MoO_3			1090		Buckley (1971)

[a] Data on shear strength are scarce, and the tensile strength data are used due to their abundance and the assertion that tensile strength of cast irons is within 20% of the shear strength (Stefanescu, 1990).
[b] Interpolated to 900°C.
[c] At room temperature.

TABLE D.2

Physical Properties of Iron and Its Oxides at 400°C

Rolls, Oxides	Density ($g\ cm^{-3}$)[c]	Hardness (HV)	Hardness (Brinell)	Latent Heat of Fusion ($cal\ g^{-1}$)	Melting Temperature (K)	Shear Strength According to Equation (9.2)
Rolls	7.7	—	400 HiCr, ICDP[h] 600 HSS[h]	23–33[k]	1500[h]	54.2
FeO	5.7[d]	248[a]	247[b]	80[f]	1650[f]	58.2
Fe_3O_4	5.18[e]	744[a]	600[b]	142.5[f]	1870[f]	107.2
Fe_2O_3	5[g]	1168[a]	760[b]	112[l]	1730[a] 1840[j]	81.8

[a] Lundberg and Gustafsson (1994).
[b] Converted from HV following Pollok (2008).
[c] Presumably at room temperature.
[d] Abuluwefa (1992).
[e] Morris, Geiger, and Fine (2012).
[f] Steinberg and Dang (1978).
[g] Gupta and Yan (2006).
[h] Sorano, Oda, and Zuccarelli (2004).
[i] Collins (2002).
[j] Rybacki et al. (2004).
[k] Cast iron (Carvill, 1994).
[l] Leth-Miller et al. (2003).

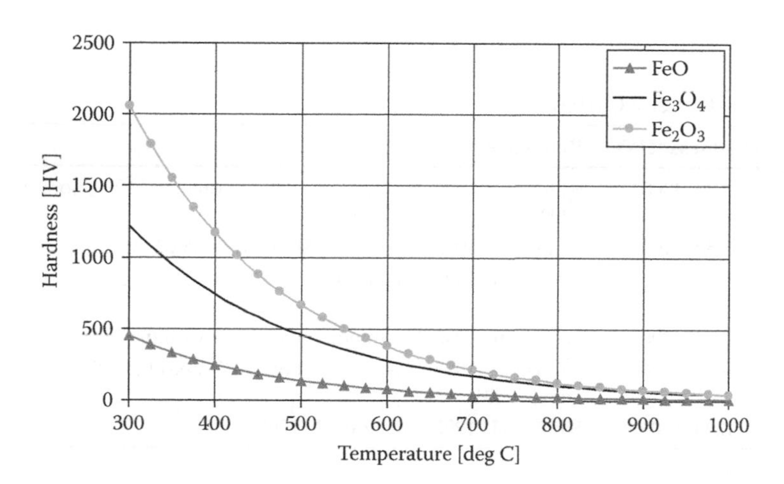

FIGURE D.1
Effect of temperature on hardness of iron oxides. (After S.E. Lundeberg and T. Gustafsson, *J. Mat. Proc. Tech.* 42:239–291, 1994. With permission.)

TABLE D.3

Physical Properties of Specific Carbides

Oxide	Hardness	Shear Strength (MPa)	Reference
Cr_3C_2	550–900 VPN	—	Mann and Prakash (2000)
Mo_2C	1460–1950 kg mm^{-2}	49	Brainard and Wheeler (1978)
Cr_3C_2	Hv 1300 kg mm^{-2}	—	Edwards et al. (1990)
Cr_3C_4	2650 kg mm^{-2}	—	Sliney (1991)
WC	2050 kg mm^{-2}	—	Sliney (1991)

TABLE D.4

Hardness of Generic Carbides

	Hardness (HV)	References
MC	2080–3300	Sano et al., 1992; Breyer et al., 1997; Byun et al., 1999; Sorano et al., 2004; Belzunce et al., 2004
M_2C	1550–2400	Sano et al., 1992; Lee et al., 2001; Collins, 2002; Belzunce et al., 2004
M_3C	800–1600	Sano et al., 1992; Hashimoto et al., 1995; Collins, 2002; Pellizzari et al., 2005
M_6C	1200–2300	Sano et al., 1992; Hashimoto et al., 1995; Collins, 2002
M_7C_3	1050–2800	Sano et al., 1992; Hashimoto et al., 1995; Lee et al., 2001; Belzunce et al., 2004
$M_{23}C_6$	1200–1800	Sano et al., 1992; Breyer et al., 1997

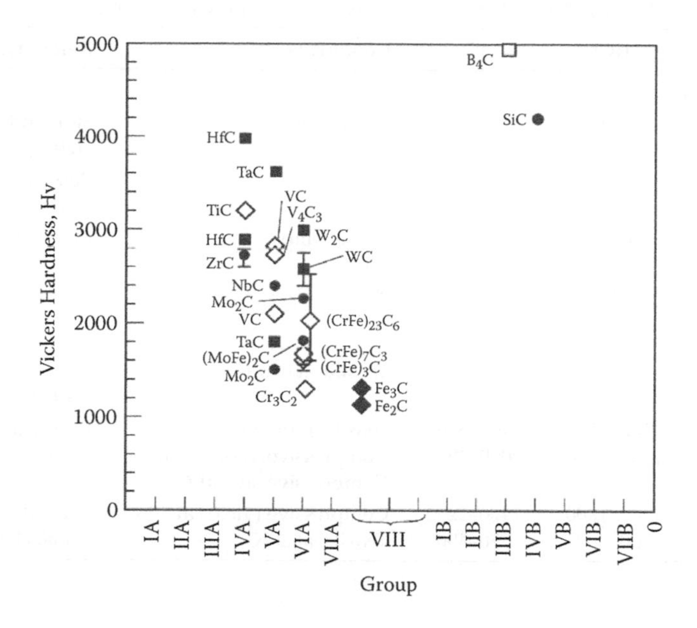

FIGURE D.2
Comparative hardness of various carbides, presumably at room temperature. (Reprinted from H. Miyahara, S.V. Bravo, K. Yamamoto et al., *ISIJ Int.* 49:1075–1079, 2009. With permission from the Iron and Steel Institute of Japan.)

TABLE D.5

Composition (in wt-%) of Key Carbides in Roll Shell

	Fe	W	Mo	Cr	V	References
MC	2–7	9–30	4.8–-25.5	3–8.2	40–73.5	Walmag et al., 2001; Lee et al., 2001; Boccalini and Sinatora, 2002; Luan et al., 2010
M_2C	4–15	10–40	26.5–60.9	8–23.3	7–15.9	Boccalini and Sinatora, 2002; Collins, 2002; Zhang et al., 2007; Luan et al., 2010
M_3C	70.6	0.2	3.7	9.8	3.1	Zhang et al., 2007
M_6C	36	35	20–33.3	4–8.5	3–11.4	Collins, 2002; Garza-Montes-de-Oca et al., 2011
M_7C_3	40–50	4–8	5–10	20–36.9	4–12.7	Lee et al., 2001; Boccalini and Sinatora, 2002; Garza-Montes-de-Oca et al., 2011

TABLE D.6
Hardness and Strength of HSS, HiCr and IC (Indefinite Chill) Rolls

HSS	HiCr	IC	Comments	Reference
Hardness (HV)				
766	524		Room temp.	Stott and Jordan (2001)
710–800	590–730	650–775	?	Belzunce et al., (2004)
884	672		Room temp.	Milan et al. (2005)
883	667		500°C	
619	614		600°C	
Strength (MPa)				
700–850	550		Tensile, at 600°C	Caithness et al. (1999)
800–1000	700–850	400–500	Tensile, room temp.	Belzunce et al., (2004)
2700–3000	1700–2200	2000–2500	Compressive, room temp.	
2450–2550			Compressive, at 500°C	
1900–2100	1600–1800	1400–2200	Compressive (yield) at 20°C	Lecomte-Beckers et al. (1997)
1000–1100	700–800	350–450	Tensile, at 20°C	

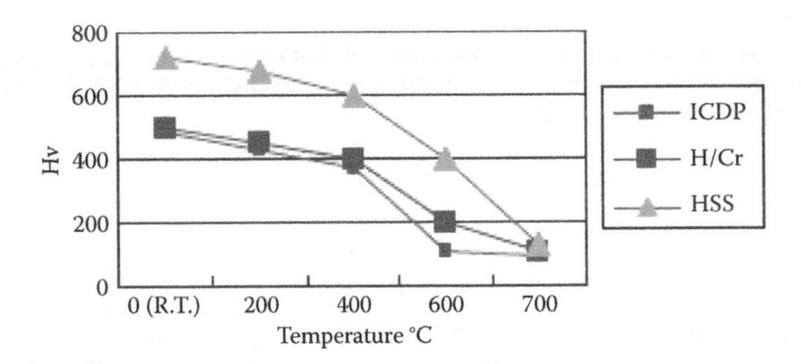

FIGURE D.3
Dependence of roll matrix hardness on temperature for various roll types. (From Sorano, H., N. Oda, and J.P. Zuccarelli, 2004, History of high-speed steel rolls in Japan, in *Proc. MS&T Conf.*, 26–29 September 2004, New Orleans, 379–390, Warrendale: The Minerals, Metals and Materials Society. Reprinted with permission of the MS&T sponsor societies.)

References

Abuluwefa, H. 1992. *Scale Formation in a Walking-Beam Steel Reheat Furnace*. M. Eng. Thesis, McGill University.

Anon. 1990. Elevated temperature properties of ferritic steels. In *ASM Metals Handbook, Vol. 1*, 617–652. Ohio: American Society for Metals.

Belzunce, F.J., A. Ziadi, and C. Rodriguez. 2004. Structural integrity of hot strip mill rolling rolls. *Eng. Fail.Anal.* 11:789–797.

Boccalini, M. Jr, and A. Sinatora. 2002. Microstructure and wear resistance of high speed steels for rolling mill rolls. In *Proc. 6th Int. Tooling Conf., Karlstad, Vol. 1,* 425–438.

Brainard, W.A., and D.R. Wheeler. 1978. *Friction and Wear of Radiofrequency-Sputtered Borides, Silicides and Carbides*. NASA TP 1156.

Breyer, J.P., R.J. Skoczynski, and G. Walmag. 1997. High speed steel rolls in the hot strip mill. In *Proc. SARUC Conf.*, Vanderbijlpark.

Buckley, D.H. 1971. *Friction, Wear and Lubrication in Vacuum*. NASA SP 277.

Byun, G., S. Oh, C.G. Lee et al. 1999. Correlation of microstructure and microfracture mechanism of five work rolls. *Met. Mat. Trans.* 30A:234–243.

Caithness, I., S. Cox, and S. Emery. 1999. Surface behaviour of HSS in hot strip mills. In *Proc. Rolls 2000+ Advances in Mill Rolls Technology Conf.*, 111–120. Birmingham: Institute of Materials.

Carvill, J. 1994. *Mechanical Engineer's Data Handbook*. Oxford: Butterworth-Heinemann.

Collins, D. 2002. The metallurgy of high speed steel rolls. In *Rolls for Metalworking Industries*, G.E. Lee (Ed.), 83–91. Warrendale, PA: Association for Iron and Steel Technology.

Eadie, D.T., J. Kalousek, and K.C. Chiddick. 2002. The role of high positive friction (HPF) modifier in the control of short pitch corrugations and related phenomena. *Wear* 253:185–192.

Edwards, P.M., H.E. Sliney, C. DellaCorte et al. 1990. *Mechanical Strength and Thermophysical Properties of PM212: A High Temperature Self-Lubricating Powder Metallurgy Composite*. NASA TM-103694.

Garza-Montes-de-Oca, N.F., R. Colás, and W.M. Rainforth. 2011. On the damage of a work roll grade high speed steel by thermal cycling. *Eng. Fail. Anal.* 18:1576–1583.

Gupta, A., and D. Yan. 2006. *Mineral Processing Design and Operation: An Introduction*. Amsterdam: Elsevier.

Hashimoto, M., T. Kawakami, T. Oda et al. 1995. *Development and Application of High-Speed Tool Steel Rolls in Hot Strip Rolling*. Nippon Steel Tech. Rep. No. 66:82–90.

Lecomte-Beckers, J., L. Terziev, and J.P. Breyer. 1997. Graphitisation in chromium cast iron. In *Proc. 39th MWSP Conf., XXXIV,* ISS, 423–431.

Lee, J.H., J.C. Oh, J.W. Park et al. 2001. Effects of tempering temperature on wear resistance and surface roughness of a high speed steel roll. *ISIJ Int.* 41:859–865.

Leth-Miller, R., P. Glarborg, L.M. Jensen et al. 2003. Experimental investigation and modelling of heat capacity, heat of fusion and melting interval of rocks. *Thermochim. Acta* 406:129–142.

Lu, X., J. Cotter, and D.T. Eadie. 2005. Laboratory study of the tribological properties of friction modifier thin films for friction control at the wheel/rail interface. *Wear* 259:1262–1269.

Luan, Y., N. Song, Y. Bai et al. 2010. Effect of solidification rate on the morphology and distribution of eutectic carbides in centrifugal casting high-speed steel rolls. *J. Mat. Proc. Tech.* 210:536–541.

Lundberg, S.-E., and T. Gustaffson. 1994. The influence of rolling temperature on roll wear, investigated in a new high temperature test rig. *J. Mat. Proc. Tech.* 42:239–291.

Luong, L.H.S., and T. Heijkoop. 1981. The influence of scale on friction in hot metal working. Wear 71:93–102.

Mann, B.S., and B. Prakash. 2000. High temperature friction and wear characteristics of various coating materials for steam valve spindle applications. *Wear* 240:223–230.

Milan, J.C.G., M.A. Carvalho, R.R. Xavier et al. 2005. Effect of temperature, normal load and pre-oxidation on the sliding wear of multi-component ferrous alloys. *Wear* 259:412–423.

Miyahara, H., S.V. Bravo, K. Yamamoto et al. 2009. Solute concentration and carbides formation for steel milling rolls. *ISIJ Int.* 49:1075–1079.

Morris, A.E., G. Geiger, and H.A. Fine. 2012. *Handbook on Material and Energy Balance Calculations in Material Processing*. New York: John Wiley and Sons.

Pellizzari, M., A. Molinari, and G. Straffelini. 2005. Tribological behaviour of hot rolling rolls. *Wear* 259:1281–1289.

Pelizzari, M., D. Cescato, and M.G. De Flora. 2009. Hot friction and wear behaviour of high speed steel and high chromium iron for rolls. *Wear* 267:467–475.

Picqué, B. 2004. *Experimental Study and Numerical Simulation of Iron Oxide Scales Mechanical Behaviour in Hot Rolling*. PhD diss. Ecole de Mines de Paris.

Pollok, H. 2008. *Umwertung der Skalen*. Qualität und Zuverlässigkeit 4.

Rybacki, E., M. Naumann, W. Schäfer et al. 2004. Glide systems of hæmatite single crystals in deformation experiments. In *Proc. Applied Mineralogy. Developments in Science and Technology, Vol. 2*, M. Pecchio et al. (Eds.), 947–949. São Paulo.

Sano, Y., T. Hattori, and M. Haga. 1992. Characteristics of high-carbon high speed steel rolls for hot strip mill. *ISIJ Int.* 32:1194–1201.

Sliney, H.E. 1991. *Solid Lubricants*. NASA TM-103803.

Sorano, H., N. Oda, and J.P. Zuccarelli. 2004. History of high speed steel rolls in Japan. In *Proc. MS&T Conf., 26–29 September 2004, New Orleans*, 379–390. Warrendale, PA: Minerals, Metals and Materials Society.

Stefanescu, D.M. 1990. Compacted graphite iron. In *Properties and Selection: Iron, Steels and High-Performance Alloys, ASM Metals Handbook, Vol.1*, 56–70. Ohio: American Society for Metals.

Steinberg, M., and V.-D. Dang. 1978. *Hydrogen Production Using Fusion Energy and Thermochemical Cycles*. Brookhaven National Lab., BNL-24209.

Stott, F.H., and M.P. Jordan. 2001. The effects of load and substrate hardness on the development and maintenance of wear-protective layers during sliding at elevated temperatures. *Wear* 250:391–400.

Vagnard. G., and J. Manenc. 1964. Etude de la plasticite du protoxyde de fer et de l'oxyde cuivreux. *Mem. Sci. Rev. Met.* 61:768–776.

Walmag, G., R.J. Skoczynski, and J.P. Breyer. 2001. Improvement of the work roll performance on the 2050 mm hot strip mill at ISCOR Wanderbijlpark. *La Rev. Met.* 98:295–304.

Zhang, X., W. Liu, D. Sun et al. 2007. The transformation of carbides during austenization and its effect on the wear resistance of high speed steel rolls. *Met. Mat. Trans.* 38A:499–505.

Appendix E: Criteria for Coil Selection

For the large-scale analyses in Chapters 15 and 16, friction in roll gap was quantified with two measures with a proprietary formula (Chapter 17.2), and as the force applied to coils of the same grade rolled under very similar conditions (in terms of temperature, input and output thickness at individual stands, and roll speed).

The ranges of temperature and thickness were determined with a set-up model to ensure that force at any end of the range will not change by more than 3% and 5% from a standard case at stands F1–F3 and F4–F5, respectively.

It was observed in the mill that friction of the same roll pair varies over a single schedule, and between schedules. For that reason, only the coils rolled during a specific part of the schedule, as indicated by the cumulative rolled length, and within the specified range of roll diameter were analysed.

The selection criteria are given in Table E.1. Some additional considerations are as follows:

1. The most common product was selected, 2.6 ± 0.05 mm, plain 0.06 wt-% C grade.
2. The measured exit strip temperature was 875 ± 10°C, save in set 2 at F3, where it was 877 ± 10°C, and set 3 at F3, where it was 880 ± 12°C.

TABLE E.1
Data Sets Selected for the Analysis and the Criteria for Coil Selection

Set	Entry Temperature before Descaler (°C)	Roll Speed (m min^{-1})	Entry Thickness at the Stand (mm)	Exit Thickness at the Stand (mm)	Cumulative Rolled Length within a Schedule (km)	Work Roll Diameter Range (mm)
Stand F1						
1	1030±15	102±5	26.45±0.15	12.3±0.1	12.5–25	>750
Stand F2						
1	1015±15	200±10	12.5±0.1	6.5±0.05	30–45	680–740
Stand F3						
1	1015±15	320±15	6.5±0.04	4.25±0.03	>35	>700
2	1010±15	320±15	6.5±0.04	4.25±0.03	>35	700–750
3	1005±20	320±20	6.5±0.2	4.25±0.15	>40	748–767
Stand F4						
1	1035±20	410±20	4.25±0.03	3.1±0.02	>20	670–690
2	1020±20	435±20	4.25±0.03	3.11±0.02	20–90	670–692
3	1010±20	430±20	4.26±0.03	3.12±0.02	30–80	670–693
5	1010±20	435±20	4.26±0.03	3.12±0.02	55–90	>673
Stand F5						
1	1030±20	525±25	3.121±0.012	2.607±0.006	30–80	715–734
2	1015±20	530±25	3.11±0.012	2.608±0.006	30–95	<755
3	1010±20	520±25	3.13±0.012	2.607±0.006	35–100	730–750
4	1010±20	540±25	3.11±0.012	2.61±0.006	60–90	758–766

Index

O

P

R

S

T

V

W

Printed and bound by CPI Group (UK) Ltd, Croydon, CR0 4YY
18/10/2024
01776261-0001